Nature after the genome

A selection of previous *Sociological Review* Monographs

Actor Network Theory and After*
eds John Law and John Hassard
Whose Europe? The Turn Towards Democracy*
eds Dennis Smith and Sue Wright
Renewing Class Analysis*
eds Rosemary Cromptom, Fiona Devine, Mike Savage and John Scott
Reading Bourdieu on Society and Culture*
ed. Bridget Fowler
The Consumption of Mass*
eds Nick Lee and Rolland Munro
The Age of Anxiety: Conspiracy Theory and the Human Sciences*
eds Jane Parish and Martin Parker
Utopia and Organization*
ed. Martin Parker
Emotions and Sociology*
ed. Jack Barbalet
Masculinity and Men's Lifestyle Magazines*
ed. Bethan Benwell
Nature Performed: Environment, Culture and Performance*
eds Bronislaw Szerszynski, Wallace Heim and Claire Waterton
After Habermas: New Perspectives on the Public Sphere*
eds Nick Crossley and John Michael Roberts
Feminism After Bourdieu*
eds Lisa Adkins and Beverley Skeggs
Contemporary Organization Theory*
eds Campbell Jones and Rolland Munro
A New Sociology of Work*
eds Lynne Pettinger, Jane Parry, Rebecca Taylor and Miriam Glucksmann
Against Automobility*
eds Steffen Böhm, Campbell Jones, Cris Land and Matthew Paterson
Sports Mega-Events: Social Scientific Analyses of a Global Phenomenon*
eds John Horne and Wolfram Manzenreiter
Embodying Sociology: Retrospect, Progress and Prospects*
ed. Chris Shilling
Market Devices*
eds Michel Callon, Yuval Millo and Fabian Muniesa
Remembering Elites
eds Mike Savage and Karel Williams
Un/knowing Bodies
eds Joanna Latimer and Michael Schillmeier
Space Travel and Culture: From Apollo to Space Tourism
eds Martin Parker and David Bell
Nature, Society and Environmental Crisis
eds Bob Carter and Nickie Charles

*Available from John Wiley & Sons, Distribution Centre, 1 Oldlands Way, Bognor Regis, West Sussex, PO22 9SA, UK

Most earlier monographs are still available from: Caroline Baggaley, The Sociological Review, Keele University, Keele, Staffs ST5 5BG, UK; e-mail srb01@keele.ac.uk

The Sociological Review Monographs

Since 1958, *The Sociological Review* has established a tradition of publishing one or two Monographs a year on issues of general sociological interest. The Monograph is an edited book length collection of research papers which is published and distributed in association with Blackwell Publishing Ltd. Our latest Monographs have been *Nature, Society and Environmental Crisis* (edited by Bob Carter and Nickie Charles), *Space Travel & Culture: From Apollo to Space Tourism* (edited by David Bell and Martin Parker), *Un/Knowing Bodies* (edited by Joanna Latimer and Michael Schillmeier), *Remembering Elites* (edited by Mike Savage and Karel Williams), *Market Devices* (edited by Michel Callon, Yuval Millo and Fabian Muniesa), *Embodying Sociology: Retrospect, Progress and Prospects* (edited by Chris Shilling), *Sports Mega-Events: Social Scientific Analyses of a Global Phenomenon* (edited by John Horne and Wolfram Manzenreiter), *Against Automobility* (edited by Steffen Böhm, Campbell Jones, Chris Land and Matthew Paterson), *A New Sociology of Work* (edited by Lynne Pettinger, Jane Parry, Rebecca Taylor and Miriam Glucksmann), *Contemporary Organization Theory* (edited by Campbell Jones and Rolland Munro), *Feminism After Bourdieu* (edited by Lisa Adkins and Beverley Skeggs), *After Habermas: New Perspectives on the Public Sphere* (edited by Nick Crossley and John Michael Roberts). Other Monographs have been published on consumption; museums; culture and computing; death; gender and bureaucracy; sport and many other areas. We are keen to receive innovative collections of work in sociology and related disciplines with a particular emphasis on exploring empirical materials and theoretical frameworks which are currently under-developed. If you wish to discuss ideas for a Monograph then please contact the Monographs Editor, Chris Shilling, School of Social Policy, Sociology and Social Research, Cornwallis North East, University of Kent, Canterbury, Kent CT2 7NF C.Shilling@kent.ac.uk

Nature after the genome

Edited by Sarah Parry and John Dupré

Wiley-Blackwell/The Sociological Review

BLACKWELL PUBLISHING
350 Main Street, Malden, MA 02148–5020, USA
9600 Garsington Road, Oxford OX4 2DQ, UK
550 Swanston Street, Carlton, Victoria 3053, Australia

First published 2010 by Blackwell Publishing Ltd

Library of Congress Cataloging-in-Publication Data

Nature after the genome / edited by Sarah Parry, John Dupré.
p. cm.
Includes bibliographical references and index.
ISBN 978-1-4443-3396-1 (pbk.)
1. Genomics—Social aspects. 2. Nature. I. Parry, Sarah
II. Dupré, John.
QH438.7.N38 2010
660.6′5–dc22

2010020645

A catalogue record for this title is available from the British Library

Set in 10/12 Times NR MT

by Toppan Best-set Premedia Limited

Printed and bound in the United Kingdom

by Page Brothers, Norwich

The publisher's policy is to use permanent paper from mills that operate a sustainable forestry policy, and which has been manufactured from pulp processed using acid-free and elementary chlorine-free practices. Furthermore, the publisher ensures that the text paper and cover board used have met acceptable environmental accreditation standards.

For further information on Blackwell Publishing, visit our website:
http://www.blackwellpublishing.com

Contents

Acknowledgements

This volume developed as a collaborative project between elements of the ESRC (Economic and Social Research Council) Genomics Network, which also provided funding for the project. We gratefully acknowledge the support of the ESRC.

We are particularly grateful to Chris Shilling and the anonymous referee who read and commented on the manuscript; also to a number of anonymous reviewers who graciously made themselves available at short notice to comment on specific chapters. We also thank Caroline Baggaley at *The Sociological Review* for her help.

SP would like to thank colleagues in the ESRC Innogen Centre (ESRC Grant numbers: RES-145-28-1004 and RES-145-28-0002) and beyond who showed enthusiasm for this project during the early stages when it was little more than a vague idea – Dave Wield, in particular, for providing financial support and encouragement. JD's work on the project was part of the programme of the ESRC Centre for Genomics in Society (Egenis) (ESRC Grant numbers: RES-145-28-1002 and RES-145-28-0001).

Part One
Introduction

Introducing nature after the genome

Sarah Parry and John Dupré

> If human beings once knew what 'nature' was, they do so no longer. What is 'natural' is now so thoroughly entangled with what is 'social' that there can be nothing taken for granted about it any more (Beck, Giddens and Lash, 1994: vii).

Since the publication announcing the discovery of the molecular structure of DNA by Watson and Crick in 1953, teams of researchers have been working to sequence the DNA-genome of whole organisms. At the time of writing, over one thousand genomes have already been sequenced ranging from viruses and fungi to plants and animals (see Genomes OnLine Database [GOLD]). As well as those considered complete there are nearly as many ongoing, incomplete sequencing projects. In spite of genome sequencing research having been underway for nearly fifty years, it was the initiation of the Human Genome Project (HGP) in 1990 that galvanized the genomic era. Ten years later, to a media fanfare and in the presence of Prime Minister Blair and President Clinton, teams of scientists from both sides of the Atlantic announced the first full draft of the human genome.

Like the Manhattan Project over fifty years earlier, the HGP mobilized and drew in vast resources – including financial, technological, and human – and involved experts spanning the biological and informational sciences. Since the first complete genome was sequenced in 1975 (of bacteriophage φX174), the technologies for sequencing genomes have improved exponentially. Nowadays automated DNA sequencing machines can sequence millions of base pairs, the As Cs Gs and Ts that make up the genetic code, in a single day. Adding to the sense of a 'biological revolution' (see, eg, http://www.unesco.org/science/wcs/abstracts/I_9_biological.htm accessed 31 October 2009) are the ongoing 'breakthrough' (see Brown, 2000) stories as genomics continues to expand in breadth and depth: the creation of Dolly the Sheep and other mammalian clones, the development of crops and non-human animals containing genetic material from other species, and numerous 'gene for . . .' announcements to explain behavioural or physical characteristics of humans (and others), to name but a few.

As the preceding examples begin to suggest, we should not suppose that the HGP is all or most of what genomic science is about. Indeed, though, it has been extremely important in creating public awareness of the science, from the perspective of the researchers it is really a sideline. Although knowledge of the human genome sequence may eventually be of great importance in combating human disease, more immediate applications are likely to come in the applications of genomic science to plants and, above all, microbes, where ethical restrictions on the kinds of experiments that can be undertaken are less constraining.

So what, exactly, is genomic science? It is, of course, science related to the study of the genome but this, it turns out, covers a remarkably wide range of activities. One reason for this is that there are actually two rather different ways of thinking about genomes, which give rise to some very different scientific projects. Often, the genome is thought of in quite an abstract way, as a body of information. This might be as a series of nucleotides, the As, Cs, Gs and Ts familiar in representations of genomic sequence, or as a series of markers, present or absent in a particular genome. This perspective facilitates the comparison of genomes, for example in inferring evolutionary relationships, or in establishing the identity of a piece of bodily material found at a crime scene with one extracted from the body of a suspect. A second perspective applies the techniques of analytic chemistry to genomes to explore the details of their structure. Here the genome is treated as a material thing. This perspective can facilitate the insights into molecular physiology that, it is hoped, will eventually provide us with medical breakthroughs, or the genetic modification of plants and microbes that may address our needs for food and energy. Much confusion about genomics has come about through failure to distinguish these perspectives (see Barnes and Dupré, 2008, for further discussion).

A second reason for the diversity of genomic sciences is that genomics has come sometimes to stand loosely for a range of contemporary biomolecular investigations, for example those referred to more technically as proteomics, metabolomics, or metagenomics, as well as systems biology, synthetic biology, and the investigation of the medical potential of stem cells. Sometimes these fields are grouped together as postgenomic sciences, but this can be misleading as genomics, whether construed broadly or narrowly, is at its beginning rather than a thing of the past. (In the title of this book, 'after the genome' refers to a point of departure, not a finished event.) The essays in this volume draw on this diverse set of meanings of genomes, informational and material, and both narrowly and more tangentially connected to the genome itself.

In public arenas, much of genomics research is couched in promissory terms – as a future means to revolutionize medicine, agriculture, ecology and conservation (see, for example, Nelkin, Dingwall and Clarke, 2002). Some such advances have indeed been made, for example with partially pest-resistant crops, though the use of such genetically modified organisms has been extremely controversial, at least in Europe (cf. Murphy and Levidow, 2006). But no doubt most eagerly anticipated are the cures for intractable diseases in humans, constantly advertised as lying just around the corner but frustratingly failing to manifest themselves as practical therapies. More concretely, genome research has provided new insights into biodiversity as well as into the history of life, and has suggested possibilities for reorienting conservation and the management of ecologies. Some have even imagined the reintroduction of extinct plant or animal species through some combination of genetic modification and cloning techniques. However, this disparate research that falls within the very general area of genomics raises many difficult questions and, consequently, has been subjected to sometimes intense public and policy scrutiny.

Throughout the 1990s, while the HGP was in progress, social scientists and philosophers were just some of those involved in reflecting upon the implications of genomics, examining questions regarding the social context of knowledge generation and its implications for human, animal and plant life.[1] Symbolically, the transformative power of the HGP was in little doubt. How we think about ourselves, other forms of life and their/our interrelations had become, many argued, 'geneticized' (cf. Lippman, 1992, see also Nelkin and Lindee, 1995). That is, genetic explanations for physical and behavioural traits were privileged over other kinds of explanations, for example cultural or, indeed, over explanations appealing to other kinds of biological phenomena that fall outside of the genomic frame (see Barnes and Dupré, 2008). Yet, materially, what this new knowledge would entail for human and non-human life was, and continues to be, viewed with uncertainty and ambivalence.

Adding to the complexity of these issues is the epistemological shift of researchers working in the field of genomic science itself. The models of genes and genomes and their interactions with environments (understood to include everything from cytoplasm to society that is not DNA) with which researchers operate are themselves increasingly complex (cf. Barnes and Dupré, 2008; Keller, 1995). Yet, at the same time we see more than echoes of more simplistic, straightforward genetic determinism – for example in relation to race or sex – that must not be ignored (see Fausto-Sterling, 2004). Notoriously the history of genetics has been intertwined with political debates about eugenics, genetic determinism, and the biological basis of, and implications for, race, gender, and class. There is a compelling argument that deeper contemporary understanding of the nature of genes and genomes, and the way these are embedded in complex systems characterized by multidirectional causal pathways, should lead to the rejection of the simplistic pictures underlying eugenic and deterministic thinking. Nonetheless, such contentious issues have not gone away, as was made clear by the demise of the Human Genome Diversity Project, undermined by the failure to deal successfully with ethnic sensitivities (Reardon, 2005). And even as simplistic conceptions of the action of genes are being abandoned by the scientific experts, it may be that they are being enthusiastically embraced in other social spheres, and even in other parts of science. At any rate, old debates about determinism are no longer (if they ever were) so straightforward (see Petersen, 2007).

While it might appear somewhat anthropocentric to foreground the HGP, as opposed to other genome research, we do so quite self-consciously. It is important to recognize the role the HGP played in mobilizing and channelling resources into genomics, thus having ramifications across the wider area. The combination of technological and scientific developments initiated by the HGP now provides researchers in genomics with the ability to examine or compare whole genomes, identifying differences and similarities between two humans or a human and a non-human animal or even a human and a dandelion. Hence, genomics quite generally, not merely specific work on the human genome, provides 'resources for individuating humans as well as celebrating their unity' (Barnes and Dupré, 2008: 83). But with this ability arise issues of identity and classification (inter and intra 'species') along with questions of commercialization and ownership in relation to biological matter. These kinds of questions feed into wider questions about inclusion/exclusion, power and democracy: How will health or environmental risks be distributed amongst human and non-human life? It is not that these issues were irrelevant to other genome projects but more that the HGP brought such issues into sharper relief while simultaneously indicating the all-encompassing power of genomics – its implications were to touch all forms of life.

It is in this context that the issue of nature arises as significant in relation to genomics. The claim that genomics-related knowledge and practices are redefining, or even threatening, nature or the natural remains a prominent feature of public, policy and academic debates. Taking animals, plants, microbes and humans as its objects of study, genomics clearly deals with diverse 'kinds' of nature. Placing biology, and therefore nature, centre stage, genomics provides new knowledge and understandings of the natural and social worlds. But more than solely knowing the world, genomics is contributing to the proliferation of artefacts in the world; the ability to add, remove and manipulate the biological matter of organisms has resulted in the existence of new things in the world. Hence the domain of the artefactual is expanding as less and less remains entirely outside the reach of human intervention. Thus, genomics is not merely about observing and studying the biological world and providing novel ways of knowing that world, but it also involves producing that world both symbolically and materially. In doing so, genomics also remodels and reformulates the relationship between the natural and social worlds, calling the attention of social scientists to questions about the epistemological and ontological implications of developments in diverse areas of the biological and the life sciences.

But where and how does one begin when offering a narrative about nature and genomics? Nature continues to be one of the slipperiest, most complex words in the English language. Nature is the very stuff that makes up our physical world; it is a material presence. But it is also something to think with as a locus of classificatory systems for ordering the world; nature is inherently about meanings and knowledge. While recognizing the proliferation of interpretations of 'nature', Szerszynski, Heim and Waterton (2003) have offered four typologies, in which nature is conceived as materiality, as process, as a world of meanings and significance, or as something abstract. Already, over 30 years ago, Raymond Williams devoted six pages in his *Keywords: A Vocabulary of Culture and Society* (1976/1983) to 'Nature'. There is little wonder that, more recently, Inglis, Bone and Wilkie (2005) identified enough writing from the social sciences and humanities to publish a four-volume anthology *Nature: Critical concepts in the social sciences*. The topic of nature is vast, and the quantity of books and papers already penned on it is inspiring and daunting in equal measure.

In spite of (or because of) this long intellectual history of looking at nature, new empirical research on genomics provides an opportunity to look at the topic of nature afresh. Understanding how nature has been conceptualized by different actors in specific historical and cultural contexts – what Kate Soper (1995: 3) calls 'the politics of the idea of nature' – how these understandings and conceptualizations have material consequences, and how, finally, these have been reinterpreted in the light of the findings of genomics (Szerszynski, Heim and Waterton, 2003; Bowker and Star, 1999), provides an illuminating perspective on questions about the social, political and cultural implications of genomics.

In order to draw together literature on nature into the case of genomics, the following discussion introduces the reader to three crosscutting themes that frame the chapters in this volume. From there we will provide an overview of the contributions.

***

The first of these themes is the idea that how something comes to be considered un/natural is not only historically and culturally specific (see Douglas, 1966/2002) but also the consequence of the interplay between the social and material. In this sense, we start with the supposition that nature is neither static nor singular – whether viewed as a material thing in the world, as a symbol or as a classificatory resource (see Franklin, Lury and Stacey, 2000). Instead, 'nature' is created as we move from case to case, and its meaning must be understood in terms of 'the gamut of causes starting with the psychological and ending with the sociological' (Bloor, 1997: 20). Thus, we are receptive to Szerszynski, Heim and Waterton's (2003) call for more active metaphors. Building on this more dynamic understanding of nature, it is timely to reflect on the disruptions to nature (as materiality and/or meaning) engendered by genomics related knowledge and practice. Through particular instances or cases in genomics research we are, therefore, interested in capturing the dynamic interactions between the material and the symbolic involved in reconceiving our knowledge of nature. If, as this volume demonstrates, there is no all-encompassing model or set of rules for understanding the relationship between the natural and social worlds, then it is important to illustrate the specificities of nature and how these are (re)produced and (re)defined.

Rabinow's (1999) notion of 'biosociality' is instructive here in reminding us to consider genomics as an instance where nature (biological life) is not immutable but instead is being crafted according to culturally driven desires such as instrumentalization and commodification (see Calvert, this volume). A key task for the social scientist is therefore to investigate the specificities of culture that are inscribed onto nature (again, as materiality or meaning). Yet, as a number of our contributors illustrate, nature should not be naively understood as passive but instead proves to be active, or lively, in that it resists being moulded into the visions of human actors. Nevertheless, this dynamic, or active, understanding of nature in relation to genomics must not be divorced from the social, cultural and political context in which it is situated. As Hilary Rose has argued in relation to science and technology more broadly:

> . . . science and technology – despite their claims to be 'neutral', 'objective' and 'above class, race and gender' – [are], in actuality, much more like the less-than-neutral society that produce[s] them (Rose, 1987: 151).

Thus, the inscription of culture onto nature has a certain specificity that must be examined – in what way, and with implications for whom, is culture so inscribed? Readers of this volume will find illuminating contributions addressing this important question.

We have already noted that genomics is transforming the world. In many areas of genomics-related research, researchers' knowledge and representations of nature are entangled with their material interventions into it. Rather than trying to disentangle different instantiations of nature as knowledge and representation from nature as materiality, this *entanglement* is of central concern to this volume and its contributors, and how it plays out in specific cases is where much intellectual work lies. This is the second of the themes we wish to emphasize as running throughout these essays. One aspect of this entanglement is the normative or disciplining powers of nature – as knowledge and representation – for different kinds of biological life.

There is an established literature looking at the constitution of natural categories, particularly with regard to gender and the body (see, for example, Butler, 1990, 1993; Martin, 1989; Merchant, 1982; Oudshoorn, 1994; Roberts, 2007; Strathern, 1992). Similarly, environmental sociologists, through analyses of topics such as climate change, pollution or natural resources, have demonstrated how the relationship between humans and their environment is intimately related to epistemological and ontological models of nature (eg Dickens, 1996; Soper, 1995; Yearley, 1996). While these are somewhat different literatures that tend to draw on different intellectual traditions (see Castree and Braun, 1998), such work has drawn attention to how narratives of nature produce dichotomous categories of same/different, normal/abnormal, nature/culture. Hence, the elision of natural with normal and other positive meanings is well understood. Evelyn Fox Keller (2008) in particular has recently called our attention to the complex associations between these two conceptions. She begins by unpicking definitions of nature, illustrating how the prevailing view of nature in the biological sciences is one of diversity and malleability. Yet, she goes on to point out:

> What is puzzling is that, for all the diversity and plurality empirical studies oblige us to grant to biological natures, it is precisely that most modern of biological sciences, namely genetics, to which contemporary writers and readers look to ground an opposition between nature and culture (Keller, 2008: 122).

In order to address this contradiction she reorients us by asking:

> . . . what, when the natural is grounded in genetics, is not natural, what is beyond or outsides of nature, what is not genetic? (Keller, 2008: 122).

Importantly, what Keller is pointing to is that the opposite of the natural can be one of two things: the non-natural and the unnatural. Now, these are not the same. The non-natural is that which has no physical existence or is outside that which is 'spontaneously formed' in the world (Keller, 2008: 123). The unnatural, while it has a physical existence in the world and while it may even be spontaneously formed, is that which is not usual, normal or expected (Keller, 2008: 123). This is a very helpful distinction and, looking no further than contributions in this volume, we could point to relevant examples of the non-natural such as interspecies entities, genetically modified plant or animal organisms, and stem cells grown in vitro. Further, the *raison d'être* of the emergent field of synthetic biology is to create the non-natural (see Calvert, this volume). Examples of the unnatural, of that which does not conform to norms, are those defined as unhealthy or deviant animal (human or non-human) bodies and behaviours.

Finally, as the third of our cross-cutting themes, we challenge ideas of genomics as knowledge of the totality of an organism in two ways. First, as Dupré explains in his chapter, the uncritical assumption that the genome represents the discovery of the essence of an organism is entirely misguided; indeed there is not even a unique genome characteristic of each organism or kind of organism. Second, our insistence on non- or extra-biological factors such as politics, culture and expertise seeks to decentre current hegemonic discourses concerning genomes.

We have noted several times above the ways in which genomics, and earlier genetics, has been entangled with vital political controversies. This points to the greatest importance of the kind of work this book represents. Clarity about what genomic science is, and how it is situated as a social activity in a much wider social and political context, offers the best hope for forestalling the misappropriations of this area of science for often malign purposes that have been so common hitherto. We hope that this work will contribute to the emergence of a more enlightened, informed, and progressive public debate on these fundamental questions.

Organization of *Nature after the genome*

The chapters in this volume are paired, with each pair reflecting a shared theme. Both the pairings and themes were identified through an iterative process in which the volume's contributors met three times over a twelve-month period. While the disciplinary backgrounds, empirical topics and theoretical leanings are evidently diverse, the opportunity to present our embryonic ideas to the group engendered the development of a common agenda without becoming narrowly focused or prescriptive. Consequently, our collective efforts have produced an exciting volume that brings into sharp relief the benefit of creating a dialogue between intellectual traditions in the social sciences and humanities, one that we hope is just the beginning for this important topic. Hence we hope that our readers will find that this volume identifies many of the emerging issues concerning the understanding of nature in the light of genomic science, and invites further and wider debate without, however, intending to offer any sense of a final word.

We begin our tour with two chapters examining how biological entities are classified. The two contributors – John Dupré and Nicola Marks – both approach classifications as socially sustained institutions in which knowledge is inherently social. They are concerned with relations of similarity and difference and how these are produced.

John Dupré's chapter, *The polygenomic organism*, takes as its focus the boundaries of biological objects. Through examples of clones, mosaicism, chimerism, our current understandings of epigenetics/epigenomics, and symbiosis, Dupré challenges current simplistically genocentric models of life. He provides us with a more sophisticated level of description of organisms, particularly encompassing the deep interconnection, or symbiosis, between living things. As is now increasingly well-known, a functioning human organism is a symbiotic system containing a multitude of microbial cells – bacteria, archaea, and fungi – without which the whole would be seriously dysfunctional and ultimately non-viable. But beyond merely pointing to this complex symbiosis, in Chapter 2 Dupré deploys this phenomenon to question the assumption that the somewhat diverse cells containing the so-called human genome exhaust the human organism. Hence, the one genome one organism assumption – an assumption, he argues, that is a clear symptom of the excessive focus in biological thinking on the search for the discrete individual – is roundly rebuffed. Further, Dupré argues that 'it is only our choice of representation that maintains the illusion that some chemically fixed entity, the genome, can be found in all our cells.' In calling for epistemological and ontological pluralism in our understandings of organisms, he simultaneously draws our attention to a more plural understanding of the boundaries of biological objects. Dupré's chapter, grounding his analysis in contemporary biological research, offers an intervention in the widest debates about genomics and the nature and significance of the genome.

Chapter 3 nicely follows on from Dupré's chapter, but with a focus on the multiple classifications of stem cells. In *Defining stem cells*, Nicola Marks begins by introducing us to the dominant story of stem cell classification. She reminds us that when definitions or stories of nature are told by authoritative figures in public arenas some of these stabilize and seem to be true reflections of reality – given by nature – rather than social constructions. Then, drawing on a combination of interviews with scientists and other publicly available accounts, she not only unravels the dominant story, pointing to the multiple ways in which stem cells are, or might be, classified but also illustrates the particular social locations in which stories take hold or otherwise. What is interesting here are the implications of highlighting or downplaying the different characteristics of stem cells, for example their therapeutic efficacy or their ability to make lots of cells types, or only a few. Here, Marks draws attention to the wider social, political and ethical issues that are at stake when classifying stem cells according to particular criteria: the epistemic authority of particular scientists or research agendas, the autonomy of science to self-regulate, the ability to attract research funding, and even ethical authority – the constitution of un/ethical research through the definition of biological matter. Nature, in Marks's analysis of stem cell definitions, is invoked as an ordering device to de/legitimate certain practices, while the practices in turn produce particular materialities. That is, the cells that are grown *in vitro* exist in the world only by virtue of the interventions of scientists. Thus Chapter 3 underscores the contingency and political nature of particular orderings of nature – classification is not pre-determined by 'reality' – while not denying the materiality of nature.

In Part three the focus turns to an examination of how genomics is reconfiguring the gene/environment, or nature/nurture, relationship. While the case studies are rather different, both contributions are concerned to describe and analyse what different 'environments' are – what aspect of an 'environment' is brought to the fore (or not), what kinds of interventions they make possible, and the extent and form of the interdependences between environments and genes (here representing nature). Both chapters examine cases of the 'non-natural' (Keller, 2008) that are considered to be the outcome of complex gene-environment interactions.

Karen Throsby and Celia Roberts, in *Getting bigger*, combine their empirical research areas to focus on two sites of human growth and development that are exhibiting population level changes widely perceived to be problematic: precocious puberty and childhood obesity. Throsby and Roberts fruitfully use the similarities and differences between these issues as a starting point for exploring the ways in which these two intersecting / diverging processes of 'getting bigger' are conceptualized. Both cases, they show, are explained in biomedical discourses according to genetic predisposition in interaction with environments. Notably, their analysis seeks to unpack all three aspects of this: genetic explanations, environments, and gene/environment interactions. In particular, their insights illustrate that the search for and privileging of genetic explanations of precocious puberty and obesity directly shape the kinds of interventions we seek into these phenomena. Here, ideas of 'natural' (and often gendered) bodies figure highly. Like Dupré, Throsby and Roberts call for a dynamic understanding of nature (genes/genomes) – one embedded in complex social contexts. They argue against separating out and quantifying the contributions of genes. Instead, it is by connecting moral questions with physical ones, along with focussing our attention on cultural norms and practices associated with gender, class, ethnicity and other forms of body-based discrimination that we may identify sites of intervention that may have lasting impacts at both individual and population levels.

Like Throsby and Roberts, Gail Davies examines the implications of genes, environments and their interactions for understanding human behaviour and the human condition. In *Transforming behaviour*, Davies' analysis centres on the role of the mouse model and related animal experiments in neuroscience for understanding humans, focussing her analysis on two interrelated levels. First, she considers the laboratory – as a site that reveals how difficult these behavioural experiments are to standardize and repeat, in large part as a result of the ways in which human capabilities are intricately involved in the performance of the experiments. Second, she explores the environmental contributions to so-called innate behaviours through reflection on findings concerning the effects of environmental enrichment for caged animals. Davies attends to the complex interplay of animal and human agencies within different spaces, exploring the movement and meaning of behaviours as these are transformed from individual events in the laboratory to standardized data which can then be circulated widely. She discloses the often reductive assumptions underlying the material practices and scientific arguments linking human diseases and the genetically modified mice in behavioural neuroscience.

Part four, *Novelty and/in nature?* is concerned with the creation of new kinds of objects in genomics-related research – the 'non-natural' (Keller, 2008). The ability of scientists to create life that does not spontaneously occur offers a unique opportunity to explore the entanglement of researchers' representations of nature with their material interventions into it. How do understandings of nature both figure in the production of novel objects and become transformed by their production?

These questions are central to Jane Calvert's analysis of synthetic biology in Chapter 6. In *Synthetic biology*, Calvert addresses multiple dichotomies of nature – the various opposites against which 'nature' is constructed, centrally those between the natural and the social and the natural and the invented. She examines how various understandings of nature are invoked to make synthetic biology acceptable. Here synthetic biology might be seen to mimic nature, improve nature or even replace nature. As a new field of science that attempts to create new kinds of biological entities, synthetic biology, Calvert argues, can be seen as an attempt to impose the image of society on nature. Calvert's insights illustrate how pressures for engineerability, commodification and standardization in the biosciences are all pulling in the same direction: towards a reconstruction of nature which can be made to serve our purposes. These pressures can be expected to have profound consequences for the kinds of living things that are brought into the world by synthetic biology. However, as Calvert concludes, the desire for prediction and control may be thwarted by nature's apparent unruliness or recalcitrance, and nature may resist being recreated in the image of society.

In Chapter 7, *Interspecies entities and the politics of nature*, Parry observes similar entanglements between researchers' representations of nature and their material interventions into it. Like those entities created by synthetic biologists, only through technoscientific intervention in laboratories can interspecies entities come into being. They are combinations of human and non-human animal life and, as such, call our attention to meanings of humanness, animalness and their relations. Drawing on transcripts from a public event where participants discussed their views of creating interspecies entities, Parry analyses how some characteristics of interspecies entities are selected and privileged over others when classifying them. She shows how existing understandings of nature (species or humanness/animalness) shape scientific research practices while, in turn, scientific research practices shape our knowledge and representations of nature. Further, Parry goes on to argue that our understanding of nature has regulatory and material implications for human and non-human animal life. However, in spite of the potential transformation that interspecies entities might engender in our understandings of humans, animals and their relations, she concludes that interspecies entities do not undo old ways of classifying but instead undergo processes of naturalization.

Part five, *Public Natures*, brings a shift in empirical emphasis to plant life. Both chapters are oriented by Jasanoff's (2004) 'co-production', emphasizing that neither natural nor social orders can be taken as causing the other, but that each is produced alongside the other. Notably, these chapters address empirical areas that in different ways extend beyond the laboratory and address novel developments in spaces where science and publics meet. That is, the science in question extends into the public domain in, as yet, unpredictable ways. Further, both cases are promissory science and are therefore characterised by future oriented claims regarding the relationship between nature and society.

Richard Milne's chapter, *Drawing bright lines*, takes as its focus the promise of biopharming, examining the material and epistemological futures for nature embedded in narratives about this new area of scientific research. In contrast with older forms of genetic modification of plants for food, biopharming seeks to genetically modify plants to produce pharmaceutical products (eg. human insulin). The use of conventional food crops – what Milne calls 'drugs in cornflakes' – raises questions about the material and symbolic boundaries of food. Milne's analysis of biopharming echoes some aspects of Calvert's on synthetic biology, particularly when examining how understandings of nature are variously invoked by people to position biopharming as acceptable or unacceptable. Conclusions as to the sameness or difference between biopharmed plants and products and more traditional agricultural food stuffs provide insights into our understandings of nature – specifically by charting how biopharming transforms the classification of material objects from 'food' to 'non-food'. In this chapter, by drawing on data from diverse sources – publics, scientists, regulators and NGOs – Milne brings the multiplicity and diversity of discourses of nature into dialogue with one another. By doing so this chapter demonstrates biopharming to be a situated, social activity that engages members of society in a most personal aspect of our lives: health and what we eat.

Barcoding nature involves the use of a short gene sequence, taken from a standard position in the genome, to identify species. Emerging out of a purported crisis in taxonomy, as a shrinking number of expert taxonomists confront ever-growing estimates of the number of species to be identified, it is a promissory science – one that, among other things, seeks to extend participation in discovering and classifying biological life to people beyond trained specialists, to democratize taxonomy. More than this, this new way of classifying biological life involves challenging old classifications of species and the creation of new ones. In Chapter 9, *Barcoding nature*, Claire Waterton critically examines the knowledge claims of those involved in the Barcode of Life Initiative. The genomic methods for assigning specimens to species employed in this approach to taxonomy provide an opportunity to revisit debates about nature and culture, the natural and the social, as they inhere in the process of ordering and classifying the natural world. Drawing on interviews with professional taxonomists, she closely observes processes of naturalization and denaturalization that occur in the practice of classifying life.

Part six, *Theorizing Nature Through Genomics*, turns to broader questions concerning how developments in genomics are feeding into social theory, particularly theories of nature. More directly than other chapters, Richard Twine and Tim Newton reflect on recent theoretical developments in social science and humanities for understanding the relationship between the biological and the social. In turn, they question whether such theorizing is up to the task of making sense of nature vis-à-vis genomics.

Chapter 10, *Genomic natures*, sets out histories and features of trans- and post-humanisms to use their tensions as a way to think about nature in genomics. Twine argues that genomics frames nature ambivalently in ways that are both faithful to *and* undermining of Enlightenment understandings. Twine develops this argument by establishing links between genomics-related research and recent theories of trans- and post- humanism; he asks whether genomics reinforces existing notions of the human and its relationship to the natural world or renders other conceptualizations more relevant? At the same time he questions the critical edge provided by various strands of trans- and post-humanisms for analysing developments in genomics, for example as these apply to citizenship, commodification and exploitation. Indeed, Twine argues that some strands of transhumanism not only legitimate but actively pursue particular misappropriations of genomic science, notably casting bodies as capital. To explore these issues, he examines two cases: hybrid embryos and comparative farm animal genomic databases. Twine seeks to decentre the genome in a rather specific way – it is the anthropocentrism inherent in genomics knowledge and practice that he wishes to challenge. Thus, in this chapter, Twine's analysis is underpinned by a non-anthropocentric critical politics which he employs to question both various strands of genomics research and trans- and post-humanisms.

Tim Newton's chapter, *Life times*, addresses the time scales of nature, especially those associated with the biological/genomic and the social. Contrary to the prevalent anti-dualist contention in contemporary social theory, Newton argues that it is difficult to collapse biological and social temporality or assume that these zones constitute a uniform terrain. Particular attention is paid to 'molecular anthropology' and the popularization of the 'ancestral genome' in illustration of this argument. These ideas involve attempts to deploy genomic knowledge towards providing evolutionary histories. Within this general domain, Newton's main focus is on a key concept from molecular anthropology, the 'molecular clock'. In temporal terms, this concept is interesting since it supposedly is grounded on a biological constant through which we can quantify evolutionary time including the many millennia of human evolutionary history. This is an idea he questions. Newton then turns this critical analysis of genomics knowledge back on social theory to highlight the problem with an overweening anti-dualism in obscuring the possibility of difference in our perception of nature.

This book is, in a way, no more than a series of vignettes looking more or less speculatively at some of the areas in which human and non-human life can expect to be changed by the advances in genomic science. It aims to be a starting place for discussion rather than a set of conclusions. What we do hope that the book will show is just how far-reaching and radical the implications of this science are. What the various chapters show is some of the ways that both the ways we think about life, and the capacities we have for intervening in life processes, are being transformed. We say that these changes are radical, though of course as commentators on the scientific developments rather than practitioners we hope that we have also managed to maintain a critical distance from our subject matter. Neither our knowledge of life nor our capacities for affecting it may be quite what the scientists tell us they are. However, it is impossible to observe the burgeoning technologies, from DNA fingerprinting to genetically modified foods, from synthetic biology to biopharming, and fail to conclude that something important is happening. We hope that this book will contribute to the rapidly developing debate about what, exactly, that is.

Note

1 Indeed, as an empirical area of investigation in the social sciences and humanities, the literature is already too vast to review here (but we would point the reader to Atkinson *et al.* (2008) *The Handbook of Genetics and Society: Mapping the New Genomic Era*).

References

Atkinson, P., P. Glasner and M. Lock, (eds), *The Handbook of Genetics and Society Mapping the New Genomic Era*, London: Routledge.

Barnes, B. and J. Dupré, (2008), *Genomes and What To Make Of Them*, Chicago: University of Chicago Press.

Beck, Ulrich, Giddens, Anthony and Lash, Scott, (1994), *Reflexive Modernization: Politics, Tradition and Aesthetics in the Modern Social Order*, Cambridge: Polity Press.

Bloor, D., (1997), *Wittgenstein: Rules and Institutions*, London: Routledge.

Bowker, G.C., and S.L. Star, (1999), *Sorting Things Out: Classification and its consequences*, London: MIT Press.

Brown, N., (2000), 'Organising/Disorganising the Breakthrough Motif: Dolly the Cloned Ewe Meets Astrid the Hybrid Pig', in Brown, N., B. Rappert and A. Webster (eds), *Contested Futures: A Sociology of Prospective Techno-Science*, Aldershot: Ashgate.

Butler, J., (1990), *Gender Trouble: Feminism and the Subversion of Identity*, London: Routledge.
Butler, J., (1993), *Bodies That Matter: On the Discursive Limits of 'Sex'*, London: Routledge.
Castree, N. and B. Braun, (1998), 'The Construction of Nature and the Nature of Construction: Analytical and Political Tools for Building Survival Futures', in Braun, B. and N. Castree (eds), *Remaking Reality: Nature at the Millennium*, New York: Routledge.
Dickens, P., (1996), *Reconstructing Nature: Alienation, Emancipation and the Division of Labour*, London: Routledge.
Douglas, M., (1966/2002), *Purity and Danger*, London: Routledge.
Fausto-Sterling, A., (2004), 'Refashioning Race: DNA and the Politics of Health Care' *Differences*, 15(5): 2–37.
Franklin, Sarah, Lury, Celia and Stacey, Jackie, (eds), (2000), *Global Nature, Global Culture*, London: Sage.
Genomes OnLine Database v.3 (GOLD), http://genomesonline.org/index.htm, (accessed 30 Oct 2009).
Inglis, D., Bone, J. and Wilkie, R., (2005), *Nature: Critical Concepts in the Social Sciences*, London: Routledge.
Jasanoff, S., (ed.), (2004), *States of Knowledge: The Co-Production of Science and Social Order*, London: Routledge.
Keller, E.F., (1995), *Refiguring Life: Metaphors of Twentieth-Century Biology*, New York: Columbia University Press.
Keller, E.F., (2008), 'Nature and the Natural', *Biosocieties*, 3: 11–124.
Lippman, A., (1992), Led (astray) by Genetic Maps: The Cartography of the Human Genome and Health Care, *Social Science and Medicine*, Vol. 35, No. 12: 1469–1479.
Martin, E., (1989), *The Woman in the Body: A Cultural Analysis of Reproduction*, Milton Keynes: Open University Press.
Merchant, C., (1982), *The Death of Nature: Women, Ecology and the Scientific Revolution*, London: Wildwood House.
Murphy, J. and L. Levidow, (2006), *Governing the Transatlantic Conflict over Agricultural Biotechnology: Contending Coalitions, Trade Liberalisation and Standard Setting*, London: Routledge.
Nelkin, D. and S.M. Lindee, (1995), *The DNA Mystique: The Gene as a Cultural Icon*, New York, W.H. Freeman and Co.
Nelkin, B., R. Dingwall and D.D. Clarke, (2002), 'The Book of Life: How the Completion of the Human Genome Project was Revealed to the Public', *Health*, 6(4): 445–469.
Oudshoorn, N., (1994), *Beyond the Natural Body: An Archaeology of Sex Hormones*, London: Routledge.
Petersen, A., (2007), 'Is the New Genetics Eugenic? Interpreting the Past, Envisioning the Future', *New Formations*, 60: 79–88.
Rabinow, P., (1999), 'Artificiality and Enlightenment: From Sociobiology to Biosociality' in Biagioli, M. (ed.), *The Science Studies Reader*, London: Routledge: 407–416.
Reardon, J., (2005), *Race to the Finish: Identity and Governance in an Age of Genomics*, Princeton, NJ: Princeton University.
Roberts, C., (2007), *Messengers of Sex: Hormones, Biomedicine and Feminism*, Cambridge: Cambridge University Press.
Rose, H., (1987), 'Victorian Values in the Test-tube: The Politics of Reproductive Science and Technology' in Stanworth, M., (ed.), (1987), *Reproductive Technologies: Gender, Motherhood and Medicine*, Oxford: Polity Press.
Soper, K., (1995), *What is Nature? Culture, Politics and the Non-Human*, Oxford: Blackwell.
Strathern, M., (1992), *Reproducing the Future: Anthopology, Kinship and the New Reproductive Technologies*, Manchester: Manchester University Press.
Szerszynski, B., W. Heim and C. Waterton, (eds), (2003), *Nature Performed: Environment, Culture and Performance*, Oxford: Blackwell/Sociological Review.
Williams, R., (1976/1983), *Keywords: A Vocabulary of Culture and Society*, London: Fontana.
Yearley, S., (1996), *Sociology, Environmentalism, Globalization: Reinventing the Globe*, London: Sage.

Part Two
Classifying Biological Entities: Epistemologies of Life

The polygenomic organism

John Dupré

Introduction: genomes and organisms

Criticisms of the excessive attention on the powers of genes, 'genocentrism', have been common for many years[1]. While genes, genomes, or more generally DNA are certainly seen as playing a fundamental and even unique role in the functioning of living things, it is increasingly understood that this role can only be properly appreciated when adequate attention is also paid to substances or structures in interaction with which, and only in interaction with which, DNA can exhibit its remarkable powers. Criticisms of genocentrism are sometimes understood as addressing the idea that the genome should be seen as the essence of an organism, the thing or feature that makes that organism what it is. But despite the general decline not only of this idea, but of essentialism in general,[2] the assumptions that there is a special relation between an organism and its distinctive genome, and that this is a one-to-one relation, remain largely intact.

The general idea just described might be understood as relating either to types of organisms or to individual organisms. The genome is related to types of organism by attempts to find within it the essence of a species or other biological kind. This is a natural, if perhaps naïve, interpretation of the idea of the species 'barcode', the use of particular bits of DNA sequence to define or identify species membership. But in this paper I am interested rather in the relation sometimes thought to hold between genomes of a certain type and an individual organism. This need not be an explicitly essentialist thesis, merely the simple factual belief that the cells that make up an organism all, as a matter of fact, have in common the inclusion of a genome, and the genomes in these cells are, barring the odd collision with a cosmic ray or other unusual accident, identical. It might as well be said right away that the organisms motivating this thesis are large multicellular organisms, and perhaps even primarily animals. I shall not be concerned, for instance, with the fungi that form networks of hyphae connecting the roots of plants, and are hosts to multiple distinct genomes apparently capable of moving around this network (Sanders, 2002). I should perhaps apologise for this narrow focus. Elsewhere I have criticized philosophers of biology and others for a myopic focus on a quite unusual type of organism, the multicellular animal (O'Malley and Dupré, 2007). Nonetheless it is unsurprising that we should have a particular interest in the class of organisms to which we ourselves belong, and this is undoubtedly an interesting kind of organism. And in the end, if my argument is successful for multicellular animals it will apply all the more easily to other, less familiar, forms of life.

At any rate, it is an increasingly familiar idea that we, say, have such a characteristic genome in each cell of our body, and that this genome is something unique and distinctive to each of us. It is even more familiar that there is something, 'the human genome', which is common to all of us, although, in light of the first point, it will be clear that this is not exactly the same from one person to another. The first point is perhaps most familiar in the context of forensic genomics, in the realization that the tiniest piece of corporeal material that any of us leaves lying around can be unequivocally traced back to us as its certain source. At any rate, what I aim to demonstrate in this paper is that this assumption of individual genetic homogeneity is highly misleading, and indeed is symptomatic of a cluster of misunderstandings about the nature of the biological systems we denominate as organisms.

Organisms and clones

A clone, outside Star Wars style science fiction, is a group of cells originating from a particular ancestral cell through a series of cell divisions. The reason we suppose the cells in a human body to share the same genome is that we think of the human body as, in this sense, a clone: it consists of a very large group of cells derived by cell divisions from an originating zygote. A familiar complication is that if I have a monozygotic ('identical') twin, then my twin will be part of the same clone as myself. Although this is only an occasional problem for the human case, in other parts of biology it can be much more significant. Lots of organisms reproduce asexually and the very expression 'asexual reproduction' is close to an oxymoron if we associate biological individuals with clones. For asexual reproduction is basically no more than cell division, and cell division is the growth of a clone. If reproduction is the production of a new individual it cannot also be the growth of a pre-existing individual. Indeed what justifies taking the formation of a zygote as the initiation of a new organism, reproduction rather than growth, is that it is the beginning of a clone of distinctive cells with a novel genome formed through the well-known mixture between parts of the paternal and the maternal genomes.

As I have noted, it is common to think of genomes as standing in one-to-one relations with organisms. My genome, for instance, is almost surely unique and it, or something very close to it, can be found in every cell in my body. Or so, anyhow, the standard story goes. The existence of clones that do not conform to the simple standard story provides an immediate and familiar complication for the uniqueness part of this relation. If I had a monozygotic ('identical') twin, then there would be two organisms whose cells contained (almost) exactly the same genomes; we would both, since originating from the same lineage-founding zygote, be parts of the same clone. And lots of organisms reproduce asexually all or some of the time, so this difficulty is far from esoteric.

Some biologists, especially botanists, have bitten the bullet here. They distinguish ramets and genets, where the genet is the sum total of all the organisms in a clone, whereas the ramet is the more familiar individual (Harper, 1977). Thus a grove of trees propagated by root suckers, such as are commonly formed, for instance, by the quaking aspen (*Populus tremuloides*), in the deserts of the South West United States, is one genet but a large number of ramets. Along similar lines it has famously been suggested that among the largest organisms are fungi of the genus *Armillaria*, the familiar honey fungi (Smith *et al.*, 1992). A famous example is an individual of the species *Armillaria ostoyae* in the Malheur National Forest in Oregon that was found to cover 8.9 km^2 (2,200 acres).[3] To the average mushroom collector a single mushroom is an organism, and it would be strange indeed to claim that two mushrooms collected miles apart were parts of the same organism. There is nothing wrong with the idea that for important theoretical purposes this counterintuitive conception may be the right one; there is also nothing wrong with the more familiar concept of a mushroom. The simple but important moral is just that we should be pluralistic about how we divide the biological world into individuals: different purposes may dictate different ways of carving things up.

It's pretty clear, however, that we cannot generally admit that parts of a clone are parts of the same individual. Whether or not there are technical contexts for which it is appropriate, I doubt whether there are many interesting purposes for which two monozygotic human twins should be counted as two halves of one organism. Or anyhow, there are certainly interesting purposes for which they must be counted as distinct organisms, including almost all the regular interests of human life. An obvious reason for this is that most of the career of my monozygotic twin (if I had one) would be quite distinct from my own. And for reasons some of which should become clearer in light of the discussion below of epigenetics, the characteristics of monozygotic twins tend to diverge increasingly as time passes. The careers of monozygotic twins may carry on independently from birth in complete ignorance of one another; but it is hardly plausible that if I were now to discover that I had a monozygotic twin, this would drastically change my sense of who I was (ie a spatially discontinuous rather than spatially connected entity). Some kind of continuing connection seems needed even to make sense of the idea that these could be parts of the same thing. Being parts of the same clone is at any rate not a sufficient condition for being parts of the same biological individual.

However, we should not immediately assume that the concept of a genet encompassing a large number of ramets is generally indefensible. A better conclusion to draw is that theoretical considerations are insufficient to determine unequivocally the boundaries of biological objects. Sometimes, perhaps always, this must be done relative to a purpose. There are many purposes for which we distinguish human individuals and for the great majority of these it would make no sense to consider my twin and myself part of the same entity. My twin will not be liable to pay my debts or care for my children, for instance, if I should default on these responsibilities, though it is interesting in the latter case that standard techniques for determining that they are my children would not distinguish my paternity from my twin's. This may even point to an evolutionary perspective from which we are best treated as a single individual. And when it comes to the trees, this is surely the right way to go. For the purposes of some kinds of evolutionary theory the single genet may be the right individual to distinguish, but if one is interested in woodland ecology, what matters will be the number of ramets. If this seems an implausible move, this is presumably because of the seemingly self-evident individuality of many biological entities. I hope that some of the considerations that follow will help to make this individuality a lot less self-evident than it might appear at first sight. But whether or not the pluralism I have suggested for individual boundaries is defensible, the assumption of a one-to-one relation between genomes and organisms is not. I will explain the objections to this assumption in what I take to be an order of increasing fundamentality. At any rate, as the next section will demonstrate, the various phenomena of genetic mosaicism suffice to demonstrate that genotypes will not serve to demarcate the boundaries of biological individuals. Or in other words, genomic identity is not a necessary condition for being part of the same biological individual.

Genomic chimeras and mosaics

The general rubric of genomic mosaicism encompasses a cluster of phenomena. An extreme example, sometimes distinguished more technically as chimerism, is of organisms that have resulted form the fusion of two zygotes, or fertilised eggs, in utero. The consequence of this is that different parts of the organism will have different genomes – the organism is a mosaic of cells of the two different genomic types from which it originated. A tragic consequence of this has been the number of cases of women who have been denied custody of their children on the basis of genetic tests that appeared to show that they and the child were not related. It has turned out that the explanation of this apparent contradiction of a connection of which the mother was in no doubt was that she was a genomic mosaic of this kind, and the cells tested to establish the parental relation were from a different origin from the gametes that gave rise to the child (Yu *et al.*, 2002). With the exception of a modest degree of chimerism found in some fraternal twins who have exchanged blood and blood cell precursors in utero and continue to have distinct genotypes in adult blood cells, such cases are generally assumed to be very rare in humans. However, chimeras do not necessarily experience any unusual symptoms, so the prevalence of full chimerism, chimerism derived from multiple zygotes, is not really known, and may be much higher than suspected.

Probably more common than chimerisms resulting from the fusion of two zygotes are those resulting from mutations at some early stage of cell division. One well-known example of this is XY Turner syndrome, in which the individual is a mixture of cells with the normal XY karyotype, the complement of sex chromosomes found in most males, and XO cells, in which there is no Y chromosome and only one X chromosome (Edwards, 1971). Turner syndrome is a condition of girls in which all the cells are XO (ie with one X chromosome missing, as opposed to the standard XX); people with XY Turner syndrome generally have normal male phenotypes, though a small percentage are female and a similar small percentage are intersexed. The large majority of fetuses with either condition are spontaneously aborted. The phenotype displayed by XY Turner cases is presumably dependent on exactly when in development the loss of the X chromosome occurs.

Chimerism is quite common in some other organisms. When cows have twins there is usually some degree of shared fetal circulation, and both twins become partially chimeric. This has been familiar from antiquity in the phenomenon of freemartinism, freemartins being the sterile female cattle that have been known since the 18th century invariably to have a male twin. This is the normal outcome for mixed sex bovine twins, with the female twin being masculinised by hormones deriving from the male twin.[4] This has occasionally been observed in other domesticated animals. Even more than for the human case, the prevalence of this, and other forms of chimerism, in nature is not known.

The chimeras mentioned so far are all naturally occurring phenomena. Much more attention has lately been attracted by the possibility of artificially producing chimeras in the laboratory. And unsurprisingly, the most attention had been focused on the possibility of producing chimeras, or hybrids, that are in part human. Recent controversy has focused on the ethical acceptability of generating hybrid embryos for research purposes by transplanting a human nucleus into the egg cell of an animal of another species, usually a cow.[5] Since all the nuclear DNA in such a hybrid is human, it can be argued that this is not a chimera at all, at least in the genetic sense under consideration. On the other hand such cells will contain non-human DNA in the mitochondria, the extra-nuclear structures in the cell that provide the energy for cellular processes.[6] No doubt the mixture of living material from humans and non-humans is disturbing to many whether or not the material in question is genetic, as is clear from controversy over the possibility of xenotransplantation, use of other animals to provide replacement organs. But this will not be my concern in the present paper (though see Parry's and Twine's papers in this volume).

Modern laboratories, at any rate, are well able to produce chimeric organisms. At the more exotic end of such products, and certainly chimeras, are such things as 'geeps', produced by fusing a sheep embryo with a goat embryo. The adults that develop from these fused embryos are visibly and bizarrely chimeric, having sheep wool on parts of their bodies and goat hair on others. Much more significant, however, are the transgenic organisms that have caused widespread public discomfort in the context of genetically modified (GM) foods (see Milne, this volume). These are often seen as some kind of violation of the natural order, the mixing together of things that Nature or God intended to keep apart (see Barnes and Dupré, 2008). Whatever other objections there may be to the production of GM organisms, it will become increasingly clear that this is not one with which I am sympathetic: organisms do not naturally display the genetic purity that this concern seems to assume.

The chimeric organisms discussed so far in this section have been organisms originating to some degree from two distinct zygotes. (The exception is the XY Turner syndrome, which should strictly have been considered in the context of the following discussion.) Other cases relevant to the general topic of intraorganismic genomic diversity, but generally referred to by the term mosaicism rather than chimerism, exhibit genomic diversity but deriving from a single zygotic origin. Such mosaicism is undoubtedly very common. One extremely widespread instance is the mosaicism common to most or all female mammals that results from the expression of different X chromosomes in different somatic cells. In the human female, one of the X chromosomes is condensed into a cellular object referred to as the Barr body and is largely inert. Different parts of the body may have different X chromosomes inactive, implying that they have different active genotypes. This phenomenon will apply to most sexually reproducing organisms, though in some groups of organisms, for example birds, it is the male rather than the female that is liable to exhibit this kind of mosaicism.[7] The most familiar phenotypic consequence of this phenomenon is that exhibited by tortoiseshell or calico cats, in which the different coat colours reflect the inactivation of different X chromosomes. Although there are very rare cases of male calico cats, these appear to be the result of chromosomal anomaly (XXY karyotype), chimerism, or mosaicism in which the XXY karyotype appears as a mutation during development (Centerwall and Benirschke, 1973).

Returning to chimerism, mosaicism deriving from distinct zygotes, a quite different but very widespread variety is exhibited by females, including women, after they have borne children, and is the result of a small degree of genomic intermixing of the maternal and offspring genomes. Though scientists have been aware of this phenomenon for several decades it has recently been the focus of increased attention for several reasons. For example, recent work suggests that the transfer of maternal cells to the fetus may be important in training the latter's immune system (Mold *et al.*, 2008). Another reason for increasing interest in this topic is the fact that it opens up the possibility of genetic testing of the fetus using only maternal blood, and thus avoiding the risks inherent in invasive techniques for fetal testing such as amniocentesis (Lo, 2000; Benn and Chapman, 2009). It should also be noted that maternal cells appear to persist in the offspring and vice versa long after birth, suggesting that we are all to some degree genomic mosaics incorporating elements from our mothers and, for women, our offspring.

One final cause of chimerism that must be mentioned is the artificial kind created by transplant medicine, including blood transfusions. Very likely this will continue to become more common as techniques of transplantation become more refined and successful. A possibility increasingly under discussion is that this will eventually be extended, through the development of xenotransplantation, to include interspecific mosaicism. At any rate, any kind of transplantation, except that involving cells produced by the recipient himself or herself, will produce some genomic chimerism. So, in summary, both natural and artificial processes, but most commonly the former, generate significant degrees of chimerism in many, perhaps almost all, multicellular organisms including ourselves. The assumption that all the cells in a multicellular organism share the same genome is therefore seriously simplistic and, as mentioned above, conclusions drawn from this simplistic assumption, for example about the violation of Nature involved in producing artificial chimeras are, to the extent that they rely on this assumption, ungrounded.

Epigenetics

The topics of chimerism and mosaicism so far discussed address the extent to which the cells that make up a body are genomically uniform in the sense of containing the same DNA sequences. This discussion runs a risk of seeming to take for granted the widely held view that, given a certain common genome, understood as a genome with a particular sequence of nucleotides (the As Cs Gs and Ts familiar to everyone in representations of DNA sequence), the behaviour of other levels of biological organisation will be determined. Perhaps a more fundamental objection to the one genome one organism doctrine is that this common assumption is entirely misguided. The reason that the previous discussion may reinforce such an erroneous notion is that the comparisons and contrasts between genomes were implicitly assumed to be based entirely on sequence comparisons. But to know what influence a genome will actually have in a particular cellular context one requires a much more detailed and nuanced description of the genome than can be given merely by sequence. And once we move to that more sophisticated level of description it becomes clear that, even within the sequence-homogeneous cell lineages often thought to constitute a multicellular organism, there is a great deal of genomic diversity. These more sophisticated descriptions are sought within the burgeoning scientific field of epigenetics, or epigenomics.

A good way of approaching the subject matter of epigenetics is to reflect on the question why, if indeed all our cells do have the same genome, they nevertheless do a variety of very different things. It is of course very familiar that not all the cells in a complex organism do the same things – they are differentiated into skin cells, liver cells, nerve cells, and so on. Part of the explanation for this is that the genome itself is modified during development, a process studied under the rubric of epigenetics or epigenomics.[8] The best-known such modification is methylation, in which a cytosine molecule in the DNA sequence is converted to 5-methyl-cytosine, a small chemical addition to one of the nucleotides, or bases, that makeup the DNA molecule. This has the effect of blocking transcription of the DNA sequence at particular sites in the genome. Other epigenetic modifications affect the protein core, or histones, which form part of the structure of the chromosome, and also influence whether particular lengths of DNA are transcribed into RNA. It is sometimes supposed that these are not 'real', or anyhow significant, alterations of the genome, perhaps because we still describe the genome sequence in the same way, referring to either cytosine or 5-methyl-cytosine by the letter C. But all this really shows is that the standard four letter representation of genomic sequence is an abstraction. As a matter of fact there are about 20 nucleotides that can occur in DNA sequences, and it is only our choice of representation that maintains the illusion that some chemically fixed entity, the genome, can be found in all our cells. If we were to change the representation to a more fine-grained description of chemical composition, we would find a much greater genomic diversity than is disclosed by the more abstract and familiar four letter code.

It is true that part of the value of the abstraction that treats the genome as consisting of only four nucleotides is that this does represent a very stable underlying structure. This has provided extremely useful applications that use stable genome sequence to compare or identify organisms, applications ranging from phylogenetic analysis to forensic DNA fingerprinting. Phylogenetic analysis, the investigation of evolutionary relations between kinds of organisms, here depends on the stability of genomes as they are transmitted down the generations, and DNA fingerprinting depends on the admittedly much shorter term stability of genome sequence within the life of the individual. Methylation, on the other hand, is reversible and often reversed. However over-emphasis on this stable core can be one of the most fundamental sources of misunderstanding in theoretical biology.

Such misunderstanding is sometimes expressed in the so-called Central Dogma of Molecular Biology.[9] This is generally interpreted as stating that information flows from DNA to RNA to proteins, but never in the reverse direction. I don't wish to get involved in exegesis of what important truth may be alluded to with this slogan, and still less into the vexed interpretation of the biological meaning of 'information' (Maynard Smith, 2000; Griffiths, 2001). What is no longer disputable is that causal interaction goes both in the preferred direction of the Central Dogma, and in the reverse direction. Epigenetic changes to the genome are induced by chemical interactions with the surrounding cell (typically with RNA and protein molecules). A reason why this is so important is that it points to a mechanism whereby even very distant events can eventually have an impact on the genome and its functioning. The classic demonstration of this is the work of Michael Meaney and colleagues, on ways in which maternal care can modify the development of cognitive abilities in baby rats, something which has been shown to be mediated by methylation of genomes in brain cells (Champagne and Meaney, 2006). The most recent work by this group has provided compelling reason to extrapolate these results to humans (McGowan *et al.*, 2009). Whether epigenetic research shows that genomes are diverse throughout the animal body of course depends on one's definition of 'genome' and one's criterion for counting two as the same. It needs just to be noted that if we choose a definition that, *pace* the points made in earlier sections, counts every cell as having the same genome, we will be overlooking differences that make a great difference to what the cell actually does.

Symbiosis and metaorganisms

In this section I want to make a more radical suggestion. So far I have considered the diversity of human (or other animal) cells that may be found in an individual organism; and the phenomena I have described are generally familiar ones to molecular biologists. In this section I shall propose that there are good reasons to deny the almost universal assumption that all the cells in an individual must belong to the same species. This may seem no more than tautological: if a species is a kind of organism then how can an organism incorporate parts or members of different species? The resolution of this paradox is to realise that very general terms in biology such as species or organism do not have univocal definitions: in different contexts these terms can be used in different ways. For the case of species, this is quite widely agreed among philosophers of biology today (see, eg Dupré, 2002, chs. 3, 4.) I am also inclined to argue something similar for organisms. Very roughly, I want to suggest that the organisms that are parts of evolutionary lineages are not the same things as the organisms that interact functionally with their biological and non-biological surroundings. The latter, which I take to be more fundamental, are composed of a variety of the former, which are the more traditionally conceived organisms. But before explaining this idea in more detail I need to say a bit more about the facts on which it is based. I shall introduce these with specific reference first to the human.

A functioning human organism is a symbiotic system containing a multitude of microbial cells – bacteria, archaea, and fungi – without which the whole would be seriously dysfunctional and ultimately non-viable. Most of these reside in the gut, but they are also found on the skin, and in all body cavities. In fact about 90 per cent of the cells that make up the human body belong to such microbial symbionts and, owing to their great diversity, they contribute something like 99 per cent of the genes in the human body. It was once common to think of these as little more than parasites, or at best opportunistic residents of the various vacant niches provided by the surfaces and cavities of the body. However it has become clear that, on the contrary, these symbionts are essential for the proper functioning of the human body. This has been recognized in a major project being led by the U.S. National Institutes of Health, that aims to map the whole set of genomes in human microbial communities, the Human Microbiome Project.[10]

The role of microbes in digestion is most familiar and is now even exploited by advertisers of yoghurt. But even more interesting are their roles in development and in the immune system. In organisms in which it is possible to do the relevant experiments it has turned out that genes are activated in human cells by symbiotic microbes, and vice versa (Rawls *et al.*, 2004). Hence the genomes of the human cells and the symbiotic microbes are mutually dependent. And it seems plausible that the complex microbial communities that line the surfaces of the human organism are the first lines of defence in keeping out unwanted microbes.[11] Since the immune system is often defined as the system that distinguishes self from non-self, this role makes it particularly difficult to characterize our symbiotic partners as entirely distinct from ourselves. Finally, it is worth recalling that we are not much tempted to think of the mitochondria that provide the basic power supply for all our cellular processes as distinct from ourselves. Yet these are generally agreed to be long captive bacteria that have lost the ability to survive outside the cell.

These phenomena are far from being unique to the human case, and arguably similar symbiotic arrangements apply to all multicellular animals. In the case of plants, the mediation of the metabolic relations between the plant roots and the surrounding soil is accomplished by extremely complex microbial systems involving consortia of bacteria as well as fungi whose my celial webs pass in and out of the roots, and which are suspected of transferring nutrients between diverse members of the plant community, suggesting a much larger symbiotic system (Hart *et al.*, 2003).

My colleague Maureen O'Malley and I (Dupré and O'Malley, 2009) have suggested that the most fundamental way to think of living things is as the intersection of lineages and metabolism. The point we are making is that, contrary to the idea that is fundamental to the one genome one organism idea, the biological entities that form reproducing and evolving lineages are not the same as the entities that function as wholes in wider biological contexts. Functional biological wholes, the entities that we primarily think of as organisms, are in fact cooperating assemblies of a wide variety of lineage-forming entities. In the human case, as well as what we more traditionally think of as human cell lineages, these wider wholes include a great variety of external and internal symbionts. An interesting corollary of this perspective is that although we do not wish to downplay the importance of competition in the evolution of complex systems, the role of cooperation in forming the competing wholes has been greatly underestimated. And there is a clear tendency in evolutionary history for entities that once competed to form larger aggregates that now cooperate.

Conclusion

It should be clear that there is a continuity between the phenomena I described under the heading of chimerism and mosaicism and those discussed in the preceding section. Living systems, I am arguing, are extremely diverse and opportunistic compilations of elements from many distinct sources. These include components drawn from what are normally considered members of the same species, as illustrated by many of the cases of chimerism, but also, and more fundamentally, by the collaborations between organisms of quite different species, or lineages, which have been the topic of the preceding section. All of these cases contradict the common if seldom articulated assumption of one genome, one organism.

One plausible hypothesis about the attraction of the one genome one organism assumption is that it represents an answer to the question, what is the *right* way of dividing biological reality into organisms. But, as I have argued throughout this essay, there is no unequivocal answer to this question. From the complex collaborations between the diverse elements in a cell, themselves forming in some cases (such as mitochondria) distinct lineages, through the intricate collaborations in multispecies microbial communities, to the even more complex cooperations that comprise multicellular organisms, biological entities consist of disparate elements working together. Different questions about, or interests in, this hierarchy of cooperative and competitive processes will require different distinctions and different boundaries defining individual entities. As with the more familiar question about species, in which it is quite widely agreed that different criteria of division will be needed to address different questions, so it is, I have argued with individuals. This is one of the more surprising conclusions that have emerged from the revolution in biological understanding that is gestured at by the rubric, genomics.

Returning finally to the distinctively human, the capacities that most clearly demarcate humans from other organisms – language, culture – are the capacities that derive from our increasing participation in ever more complex social wholes. A further extension of the argument sketched in the preceding paragraph would see this as the next stage in the hierarchy of collaboration and perhaps, as has often been speculated, genuinely marking the human as a novel evolutionary innovation. Rather less speculatively, it is arguably a striking irony that the often remarked centrality of individualism in the last 200 years of social theory has perhaps been the greatest obstacle to seeing the profoundly social, or anyhow cooperative, nature of life more generally.

Acknowledgement

I gratefully acknowledge support from the Economic and Social Research Council (ESRC) and the Arts and Humanities Research Council (AHRC). The research in this paper was part of the programme of the ESRC Centre for Genomics in Society (Egenis).

Notes

1 For a recent example, see Barnes and Dupré 2008.

2 At any rate among philosophers concerned with the details of scientific belief. Essentialism has had something of a resurgence among more abstractly inclined metaphysicians (Ellis, 2001; Devitt, 2008).

3 See http://www.scientificamerican.com/article.cfm?id=strange-but-true-largest-organism-is-fungus (accessed 2 Nov 2000).
4 Exactly to what extent this is the normal outcome remains as with so many phenomena in this area somewhat unclear, however (Zhang *et al.*, 1994).
5 Although research involving hybrid embryos is generally thought unacceptable unless there are clear potential medical benefits, opinion in the UK is quite finely divided on this topic (Jones, 2009).
6 As a matter of fact the mitochondria are now known to be descendants of bacteria that long ago became symbiotically linked to the cells of all eukaryotes, or 'higher' organisms. This may suggest a further sense in which we are all chimeric, a suggestion I shall elaborate shortly.
7 Curiously, however, it appears that birds find less need to compensate for the overexpression of genes on the chromosome of which one sex has two (in birds the male has two Z chromosomes). So this kind of mosaicism will be less common, or may not occur at all (Marshall Graves and Disteche, 2007).
8 It appears that the phenomenon in question may not be fully explicable at all, however, as gene expression is also importantly affected by random processes, or noise (Raser and O'Shea, 2005). But there is also growing evidence that noise of this kind may be adaptive, and hence this effect may have been subject to natural selection (Maamar *et al.*, 2007).
9 This phrase was introduced originally by Francis Crick, and I have no wish to accuse Crick himself of misunderstanding. Indeed the use of the word 'dogma' suggests a degree of irony.
10 See http://nihroadmap.nih.gov/hmp/ (accessed 28 Oct 2009).
11 More traditional views of the limits of the human organism might make it seem strange that a strong correlate of infection with the hospital superbug, *Clostridium difficile* is exposure to powerful courses of antibiotics, though this correlation is not quite as pervasive as was earlier thought (Dial *et al.*, 2008).

References

Barnes, B. and J. Dupré, (2008), *Genomes and What to Make of Them*, Chicago: University of Chicago Press.

Benn, P.A. and A.R. Chapman, (2009), 'Practical and ethical considerations of noninvasive prenatal diagnosis', *JAMA*, 301(20): 2154–2156.

Centerwall, W.R. and K. Benirschke, (1973), 'Male tortoiseshell and calico (T – C) cats', *The Journal of Heredity*, 64: 272–278.

Champagne, F.A. and M.J. Meaney, (2006), 'Stress during gestation alters postpartum maternal care and the development of the offspring in a rodent model', *Biological Psychiatry*, 59: 1227–1235.

Devitt, M., (2008), 'Resurrecting biological essentialism', *Philosophy of Science*, 75: 344–382.

Dial, S., A. Kezouh, A. Dascal, A. Barkun and S. Suissa, (2008), 'Patterns of antibiotic use and risk of hospital admission because of *Clostridium difficile* infection', *CMAJ*, 179(8): 767–772.

Dupré, J., (2002), *Humans and Other Animals*, Oxford: Oxford University Press.

Dupré, J. and M.A. O'Malley, (2009), 'Varieties of living things: Life at the intersection of lineage and metabolism', *Philosophy and Theory in Biology*, http://www.philosophyandtheoryinbiology.org/.

Edwards, O.M., (1971), 'Masculinized Turner's syndrome XY-XO mosaicism', *Proceedings of the Royal Society of Medicine*, 64(3): 300–301.

Ellis, B., (2001), *Scientific Essentialism*, Cambridge: Cambridge University Press.

Griffiths, P., (2001), 'Genetic information: A metaphor in search of a theory', *Philosophy of Science*, 68: 394–412.

Harper, J.L., (1977), *Population Biology of Plants*, London: Academic Press.

Hart, M.H., R.J. Reader and J.N. Klironomos, (2003), 'Plant coexistence mediated by arbuscular mycorrhizal fungi', *Trends in Ecology & Evolution*, 18: 418–423.

Jones, D.A., (2009), 'What does the British public think about human–animal hybrid embryos?' *Journal of Medical Ethics*, 35: 168–170.

Kripke, S., (1980), *Naming and Necessity*, Cambridge, MA.: Harvard University Press.

Lo, Y.M.D., (2000), 'Fetal DNA in maternal plasma: Biology and diagnostic applications', *Clinical Chemistry*, 46: 1903–1906.

Maamar, H., A. Raj and D. Dubnau, (2007), 'Noise in gene expression determines cell fate in *Bacillus subtilis*', *Science*, 317: 526–529.

Marshall Graves, J.A. and C.M. Disteche, (2007), 'Does gene dosage really matter?', *Journal of Biology*, 6(1): 1.

McGowan, P.O., A. Sasaki, A.C. D'Alessio, S. Dymov, B. Labonté, M. Szyf, G. Turecki and M.J. Meaney, (2009), 'Epigenetic regulation of the glucocorticoid receptor in human brain associates with childhood abuse, *Nature Neuroscience*, 12: 342–348.

Maynard Smith, J., (2000), 'The concept of information in biology', *Philosophy of Science*, 67: 177–194.

Mold, J.E., J. Michaëlsson, T.D. Burt, M.O. Muench, K.P. Beckerman, M.P. Busch, T.-H. Lee, D.F. Nixon and J.M. McCune, (2008), 'Maternal alloantigens promote the development of tolerogenic fetal regulatory T cells in utero,' *Science*, 322: 1562–1565.

O'Malley, M.A. and J. Dupré, (2007), 'Size doesn't matter: Towards a more inclusive philosophy of biology', *Biology and Philosophy*, 22, 2007: 155–191.

Raser, J.M. and E.K. O'Shea, (2005), 'Noise in gene expression: Origins, consequences, and control', *Science*, 309: 2010–2013.

Rawls, J.F., B.S. Samuel and J.I. Gordon, (2004), 'Gnotobiotic zebrafish reveal evolutionarily conserved responses to the gut microbiota'. *PNAS*, 101(13): 4596–4601.

Sanders, I.R., (2002), 'Ecology and evolution of multigenomic Arbuscular mycorrhizal fungi', *The American Naturalist*, 160: S128–S141.

Smith, M.L., J.N. Bruhn and J.B. Anderson, (1992), 'The fungus *Armillaria* bulbosa is among the largest and oldest living organisms', *Nature*, 356: 428–431.

Yu, N., M.S. Kruskall, J.J. Yunis, H.M. Knoll, L. Uhl, S. Alosco, M. Ohashi, O. Clavijo, Z. Husain, E.J. Yunis, J.J. Yunis and E.J. Yunis, (2002), 'Disputed maternity leading to identification of tetragametic chimerism', *The New England Journal of Medicine*, 346: 1545–1552.

Zhang, T., L.C. Buoen, B.E. Seguin, G.R. Ruth and A.F. Weber, (1994), 'Diagnosis of freemartinism in cattle: the need for clinical and cytogenic evaluation', *Journal of the American Veterinary Medicine Association*, 204(10): 1672–1675.

Defining stem cells? Scientists and their classifications of nature

Nicola J. Marks

Introduction

Stem cell research (SCR) is a regular feature of news stories and public debates. In these, the existence of precisely defined, and definable, entities such as adult or embryonic stem cells is often taken for granted. This chapter examines these definitions and shows them to be fiercely contested, even within the scientific community. It argues that stem cells, and by extension nature, can be classified in particular ways and that this has epistemological, ontological and socio-political implications.

Debates within scientific communities as well as in more public settings such as parliamentary inquiries, have examined how SCR should be regulated and funded locally and internationally. During these, stem cells are described and classified in specific ways (eg Hauskeller, 2005b) which can have direct material implications. For instance, 'foetal', 'embryonic' and 'adult' SCR can be regulated by different bodies, and one kind of SCR can be legal (eg adult) whilst another can be banned (eg embryonic). Financial resources also become available to those who can define their work as SCR and can for example join prestigious research institutes endowed with large financial resources and which specifically study these cells (eg the Australian Stem Cell Centre in Melbourne or the Wellcome Trust Centre for Stem Cell Research in Cambridge). However, not all stem cell funding is available for all types of SCR, and some will only go to adult SCR (eg the National Adult Stem Cell Research Centre in Australia) and some specifically exclude research that destroys embryos (eg the European Seventh Framework Programme). Thus definitions about what counts as a stem cell or as a sub-area of SCR take on great importance.

Accounts of SCR, in particular those aimed at non-specialist audiences (eg in the media and 'outreach' documents), tend to suggest that the classifications of stem cells into specific groups (eg 'adult' and 'embryonic') derive from objective understandings of nature. In contrast, this chapter will show that 'nature', different kinds of stem cells here, can be ordered in particular ways which reflect not only the material characteristics of the entities in question, but also the goals and interests of those doing the classifying. It will also explore the social function of the *idea* of nature. This will underscore the contingency and political nature of particular orderings of 'nature', without denying the latter's materiality (Castree and Braun, 1998).

To explore classifications of nature, a useful starting point is scientists' discourses. Indeed, these often frame science-public interactions (critiqued in Nelkin, 1975; Wynne, 1996; Goven, 2006) and scientists hold great epistemic authority (Shapin, 1995). What they say is usually accepted as reflecting truth and rationality and 'anyone who would be widely believed and trusted as an interpreter of nature needs a license from the scientific community' (Barnes and Edge, 1982: 2). Therefore, this chapter will focus on what scientists say about SCR, in particular how they order stem cells.

The body of the chapter opens with a brief overview of the literature on classification and boundary work, and how nature is conceptualized here. The ontological fluidity of 'stemness' and the difficulty in defining a stem cell are then highlighted. These enable the co-existence of different stem cell stories, some of which are explored here. A stem cell story which was dominant in the UK parliament in 2000–1 is contrasted to stories and definitions given by stem cell researchers a few years later. Four dominant ways of ordering and defining stem cells are examined, where differentiation potential, therapeutic potential, safety and control, and nature itself are used as classification criteria.[1] The chapter concludes by discussing how particular stem cell stories and definitions are seen as 'ontologically robust' (Jasanoff, 2005) and therefore can temporarily stabilize and achieve epistemic success. Finally, suggestions for taking account of the contingency of nature in public discussions are offered.

Classifying and mapping out nature: some helpful literature

In order to make sense of the world, we organize things – material objects or concepts – into relations of similarity and dissimilarity; that is, we classify them. Yet, how we classify things is not pre-determined by 'reality' and instead we must understand classifications as socially sustained 'institutions' (see Barnes, 1983) underpinned by knowledge that is, itself, inherently social (see Bloor, 1976). Importantly, when we encounter a new object, we extend our classification by *analogy* and not by *identity* (Barnes *et al.*, 1996: 56). That is, new relations of similarity are not determined by previously settled relations of similarity. Thus, even though we assign particular labels to particular objects, the future use of these labels are not determined by their past uses, and future classifications are not determined by past classifications (Barnes *et al.*, 1996, especially 55–9). In fact, even if a label remains the same, its meaning can change over time and, to understand the classifications accepted in a particular community, one must examine not only the entities which are clustered, but also those who are clustering and how this can change (Barnes, 2002).

Following this line of argument – that a number of classificatory schemes are possible – it is important to ask *why* particular relations of similarity take priority over others (see Barnes, 2002). For this, it is central to consider the goals and interests of people who are classifying. Importantly though, and similar to Hauskeller, this is not to suggest that language is purely strategic; rather, it is 'diverse and multifaceted' (Hauskeller, 2005a: 40) and choices of particular labels may indicate the relevance of that scheme in that instance, rather than a desire to dupe anyone. As Barnes (2002) reminds us, one's goals and interests do not guide classifications *instead* of references to nature or reality; rather, the adequacy of a particular classificatory scheme in categorizing particular instances of nature or reality only makes sense within the context of particular goals and interests. So for example, classifying diseases according to their genetic cause makes sense, and is useful, to a group of researchers who are trying to develop genetic tests to predict the incidence of these diseases; but classifying them according to their symptoms may make more sense to doctors who are trying to treat particular symptoms individually. Neither the genetic nor the symptom-based classification is better, or more 'natural' or 'real', than the other; each does, however, serve different purposes.

To explore the implications of classifications for the cognitive authority of different actors or areas of SCR, Gieryn's concepts of 'boundary-work' and 'cultural cartography' provide us with additional conceptual tools. Moving beyond traditional essentialist criteria for the demarcation of science from non-science Gieryn (eg 1983, 1995, 1999) chooses instead to investigate how scientists erect boundaries between what they portray as 'science' and as 'non-science', and how the construction of these boundaries can contribute to scientists' authority. By introducing the concept of 'cultural cartography', which is 'a mapping out of epistemic authority, credible methods, reliable facts' (Gieryn, 1999: 4), Gieryn finds that people of opposing views draw out different representations or 'maps' of science. Each map justifies why science is special. Gieryn calls this process 'boundary-work', which is:

> the discursive attribution of selected qualities to scientists, methods, and scientific claims for the purpose of drawing a rhetorical boundary between science and some less authoritative residual non-science (Gieryn, 1999: 4–5).

As such, the boundaries of science are very fluid and permeable; they include different characteristics depending on who is drawing them and when, for which purpose and for what audience. Further, episodes of boundary-work are more visible in times of struggle, when there is competition for funding or when the authority of science is at stake.

Gieryn, like Barnes, does not focus on who is right or wrong, or which science becomes vindicated by history. Rather, boundary-work corresponds to the *rhetorical* construction, de-construction and re-construction of boundaries and is under-determined by reality (Gieryn, 1999: 18–9). Correspondingly, in considering the case of SCR our focus should not be the *correctness* of maps and classifications. As Gieryn might argue, we may learn something about 'real' stem cells by looking at these maps, and, as Barnes might argue, we may learn more about the people making the classifications: we can never *know* what stem cells *are*, but the way they are clustered indicates particular interests and promotes particular areas of SCR and ways of seeing the world.

While much of Gieryn's work has analysed how boundary-work is used to erect boundaries *around* science (ie science from non-science), it is equally applicable and illuminating to extend this analysis to examine the boundaries *within* science (as Gieryn suggests, eg 1983: 791, 1999: 34). This allows us to ask how the *best* stem cells and areas of SCR are delineated. Such questions connect us with the 'sociology of expectations' and its emphasis on discourses of promise (eg Brown and Michael, 2003; Kitzinger and Williams, 2005; Brown and Kraft, 2006; Selin, 2007). Notably, the expectations embedded in the imagined futures of scientists have been shown to influence research in the present, including the classification of nature (see also Milne, this volume).

As with 'reality' (see Barnes, 1983), nature here is seen as something that can be clustered, organized and mapped in various socially contingent ways. Nature also operates as an *idea* which can be drawn upon and utilized to give authority to particular statements or systems of classification. This has two aspects. Firstly, it relates to the use of 'nature' or 'natural' as criteria for classifying or mapping out the world; that is, why 'natural' is often synonymous with 'good'. Secondly, it relates to how and why it is important to portray one's classifications or maps as 'natural' or 'given by nature'. This becomes particularly visible in the context of public debates since:

> The trophy of public debate is to turn a particular interpretation into an accepted fact which seems beyond the stage of negotiation (Van Dyck, 1995: 13).

Often, to say that a definition is 'given by nature', or that a classification is 'natural' renders it unquestionable. In contrast, a definition that is described as steeped in culture, politics, subjectivity or ideology becomes less authoritative. As Haraway argues, '[t]he power to define what counts as technical or as political is very much at the heart of technoscience' (1997: 89).

Multiple stem cell stories, definitions and classifications

This chapter draws on qualitative data collected from different social locations. It explores in-depth interviews and group discussions undertaken in 2004–5, with 54 stem cell researchers, of various levels of seniority, working in the UK or Australia. It also looks at publicly available material (parliamentary inquiry transcripts and published scientific paper). It examines both dominant and marginal scientific voices, as well as public and private discourses.

This diversity is important because, as Jasanoff argues, issues are 'framed' in a particular way, in a particular place, at a particular time (2005: 23): local socio-cultural contexts (such as the history of a region, the actors that hold epistemic authority, the regulations already in place etc.) will shape how people make sense of experiences and phenomena. Therefore, different countries (or different communities) may interact with SCR in different ways; each will have particular frames through which science, as culture, is refracted. This has resulted in a patchwork of views and regulations relating to SCR. For example in some countries, both embryonic stem cell research (ESCR) and adult stem cell research (ASCR) are allowed and funded by the state (such as in the UK and Australia), whilst in others, only ASCR is permitted (eg Germany or France).[2]

After examining the contingency and fluidity of 'stemness' which enable different stem cell stories and definitions to co-exist, the remainder of this section explores which ones take hold in particular social locations and links this to broader socio-cultural contexts.

Infinite regress and the fluidity of stemness

Stem cells divide and give rise to other, more specialized cells, which in turn form other cells, then tissues and organisms; that is, they differentiate into different cell types. They enable lizards' tails to grow anew after being chopped off; they enable bone marrow transplants to repopulate a person's blood and rebuild their immune system; they enable skin and the gut lining to renew itself throughout adult life; and they enable tissue repair in multiple organs.

There is a spectrum of definitions for 'stem cells'. For instance, a contemporary well-respected stem cell organization, the International Society for Stem Cell Research, defines stem cells as:

> Cells that have both the capacity to self-renew (make more stem cells by cell division) as well as to differentiate into mature, specialized cells (www.isscr.org/glossary accessed 13, January 2010).

However, the defining characteristics of these cells are contested and various other stem cell definitions have been put forward. For instance, in 1978, a group of scientists, composed mainly of haematologists, described stem cells as:

> Cells with extensive self-maintaining (self-renewal capacity), extending throughout the whole (or most) of the life-span of the organism. Differentiation potential is a property of some types of stem cells but it is *not an essential feature of stem-ness* (Lajtha quoted in Lajtha, 1983, emphasis added).

In this second definition, 'differentiation potential' – that is the ability to give rise to a number of different cell types – is not the main defining characteristic of 'stemness'. Rather, it is the ability of a cell population to maintain itself over long periods of time that is defining. The latter is a property of haematopoietic (blood) stem cells. A third definition comes from a paper published in 2005: 'The Stem State: Plasticity Is Essential, Whereas Self-Renewal and Hierarchy Are Optional' (Zipori, 2005). This third way of defining stem cells focuses on a different property of these: it suggests that 'plasticity' – often used interchangeably with 'differentiation potential' – defines 'stemness'.

There could be a number of reasons to define stem cells in these different ways. The divergence could be due to different temporal contexts: whereas the second definition was given in the late 1970s, when haematopoietic stem cells were the best-known stem cell type, the third one was given more recently, when embryonic stem cell research was perhaps more dominant. This divergence could also be due to different research focuses: researchers in haematopoietic stem cells still tend to focus on self-renewal (eg Adams and Scadden, 2006: 333), whilst embryonic stem cell researchers tend to focus on differentiation potential (eg Trounson, 2006: 208). The important point here is that *defining* stem cells is a difficult task. Indeed an entire consortium has been set up to do this: the International Stem Cell Initiative (Eriksson and Webster, 2008).

Currently, stem cells are often 'defined by their not-yet-ness', by the promise of what they might divide into (Eriksson and Webster, 2008: 62), that is by their 'differentiation potential'. However, what this may be is impossible to predict with complete certainty. Indeed, part of the conceptual fluidity surrounding SCR results from the impossibility of *pointing* to a cell and saying: this is a stem cell, and it can differentiate into x other cell types. For example in humans, the differentiation potential of a group of cells is *indicated* by removing them from their biological location, creating a cell line, and injecting some of these cells into immune deficient mice, to create a tumour. Researchers consider a high differentiation potential as signified by the presence of many different kinds of cells in the tumour. Cell surface markers are used to reveal the kinds of cells present, but this is an indirect measure and to draw any conclusions, researchers need to believe that the link between cell surface markers and cell types has been accurately established. If a scientist disputes the link between a cell's differentiation potential and the surface markers of a tumour, further repeating this experiment will not convince them of its validity; rather it is through social negotiations that claims about these cells will be settled. This is what Collins and Pinch name the 'experimenter's regress' (Collins and Pinch, 1993) and it highlights that claims about stem cells' differentiation potential are 'theory-laden' and 'under-determined by reality' (Bloor, 1976). In addition, culture conditions, cell isolation protocols and other laboratory specific conditions all affect the way stem cells will behave, making them even more heterogeneous. As Eriksson and Webster argue, there is a difference between the '"public" stem cell, often represented in pictures as a stable and unproblematic entity, and its recalcitrant laboratory counterpart' (2008: 68).

Thus, it is not possible to observe cells and *know* what their differentiation potential is. Since 'nature' does not determine stem cell definitions and stories, it is important to examine how the latter are constructed and do, if temporarily, stabilize.

Story A: a dominant narrative

Different stem cells' stories resonate with particular actors and take hold in particular social locations. Some only remain relevant in small private communities, whilst others become dominant in public *fora* and directly influence policy. I have indentified one such story, reproduced below in my words:

> Story A: There are two broad categories of human stem cells: adult stem cells (AS cells) which are undifferentiated cells found in different parts of the adult body, and embryonic stem cells (ES cells) which come from embryos (usually created through *in vitro* fertilisation [IVF], initially for fertility treatments and subsequently not used for these). ES cells are more promising than AS cells because they can 'differentiate' into (give rise to) all the other cell types of the body – they are called 'pluripotent'. They therefore hold great therapeutic promise as their 'differentiation potential' indicates therapeutic potential. AS cells have a lower 'differentiation potential' and can usually only give rise to one group of cells, such as the cells needed to make blood – they are only 'multipotent'. They therefore have a lower therapeutic potential. They do have an ethical advantage: their use does not require the destruction of human embryos. However, we should not let people die by refusing to support a life saving area of research just because it destroys embryos, which are only a ball of cells and not fully human.

This story was instrumental in providing a future for SCR at the turn of the century in the UK, where scientists needed to promote ESCR, lest it become banned. Parry shows that proponents of this work needed to put embryonic stem cells forward as the best – quickest, most likely to be successful – way of obtaining therapies.[3] This involved recruiting patients and 'constructing a demand' (Parry, 2003b: 187, 2009), as well as downplaying the potential of AS cells (Kitzinger and Williams, 2005: 738; Parry, 2009) by highlighting the lower differentiation potential of these compared to ES cells (Parry, 2003b: 195). It also involved creating the link between *differentiation* potential and *therapeutic* potential, which previously had not been made (Rubin, 2008). In this particular setting, the result was continued support in the UK for research using human embryos through a change in legislation: the update of the Human Fertilisation and Embryology (HFE) Act (1990) with the introduction of the Human Fertilisation and Embryology (Research Purposes) Regulations (2001).

The above story was mobilized by a particular 'community of promise': a network of usually geographically separated actors who share views of the world and crucially also share imagined futures (Martin *et al.*, 2008). This group succeeded in enrolling others and facilitated the translation of these imagined futures and promises into a state-supported legitimate area of research. However, other stem cell stories and definitions become important in different settings and give rise to different outcomes (eg see Hauskeller *et al.*, 2005, or the special issue on this in *Science as Culture*, 2008). Some of these are now explored.

Four classification schemes

Scientists in this study cluster and draw boundaries around what they describe as good SCR or useful stem cells in four different ways, some of which differ markedly from story A above. Before going into the details, it is essential to highlight that the social contexts of these data are different to those of story A. The UK interviews took place in 2005, four years after the HFE Act was updated to allow research on human embryos for the study of 'serious diseases'; thus the legitimacy of the ESCR was no longer at stake in the policy arena.[4]

The interview settings were private and scientists did not need to defend their authority as they do in public debates (Parry, 2009). The Australian interviews took place in 2004–5, after research on 'left-over' IVF embryos was legalized, but whilst research on embryos created through cloning was still illegal.[5] Importantly, respondents in Australia reflected on the process of the public and parliamentary debates, which had taken place in 2001–2, before nationally consistent stem cell legislation was enacted.

The first way in which stem cells are classified in this study is according to their differentiation potential, from which some scientists automatically infer *therapeutic* potential: they argue that the best cells to study are those which can give rise to the most different kinds of cell types, since these will have the best range of therapeutic applications. As mentioned in story A, this was a view put forward by defenders of ESCR in the UK in 2000–1 when the criterion of 'differentiation potential' advantaged this area of research.

However, a paper was published in 2002, by Jiang and colleagues, which challenged the link between 'differentiation potential is important' and 'ES cells are better than AS cells'. It suggested that AS cells from bone marrow could 'trans-differentiate' into cells from many other tissues, such as nerve cells and skin cells, rather than only giving rise to different kinds of blood cells as previously thought. That is, they seemed to give rise to cells from a different 'lineage' to the one they belong to.[6] This raised the possibility that AS cells could have the same differentiation potential as ES cells; the idea that AS cells could be called 'pluripotent' rather than just 'multipotent' was put forward.[7] The new knowledge claims about the materiality of AS cells – their higher differentiation potential – mapped onto the interests of powerful political groups (such as Prolife movements) and challenged the story of ESCR as the best route to therapy. The criterion of differentiation potential now seemed to advantage ASCR. For example, the possibility of 'pluripotent adult stem cells' was put forward as a reason to stop ESCR by groups such as the Catholic Women's League in their submission to the Australian parliamentary inquiry.

This created a problem for those who were, at the time, fighting for the legalization of ESCR in Australia. Many scientists interviewed for this study only saw ASCR as a serious challenge to ESCR once Jiang *et al.*'s paper was published. One, Peter, said that it gave a 'scientific arm' to 'the Catholic lobby groups in particular'.[8] That is, the argument that AS cells are better than ES cells simply because they raise fewer ethical issues did not create as much concern for these scientists as the argument that AS cells are better than ES cells because they can do the same job as ES cells, *and* raise fewer ethical issues. This technical criterion was seen to carry more weight than the solely ethically based arguments raised by opponents of embryo research previously.

For some researchers, the published results were convincing and questioned the assumption that ES cells are the best route to therapy; these scientists accepted the existence of 'pluripotent AS cells'. Others however, both in parliament and during interviews, disputed the conclusions that these cells displayed as much differentiation potential as ES cells. They talked about the hype and politicization surrounding the Jiang publication. This mapped onto the broader international controversy over the existence of these cells. The idea that AS cells can 'trans-differentiate' was challenged. Instead, it was suggested they *fuse* with pre-existing pluripotent stem cells or differentiated cells (see Alison *et al*., 2003 for a review).[9] The key point here is that the critics of the publication believe that the nerve cells (for example) do not come directly from an AS cell (a bone marrow stem cell) that has transformed itself; rather they come from the fusion of a bone marrow stem cell with an existing nerve or ES cell. Therefore these AS cells lack the all important intrinsic property of being able to give rise to cells from other lineages *on their own*. They need the help of other cells, making them less useful in terms of therapy and research in the eyes of these particular scientists.

The critics of AS cell trans-differentiation challenge the 'ontological robustness' (Jasanoff, 2005: 153) of the 'pluripotent AS cell': that is, they see this label as merely 'purpose-serving language' which does not does not map onto the materiality of stem cells. By doing this, they can maintain that differentiation potential is a criterion to measure promise, but contest the changed location of AS cells in the classificatory scheme from low to high promise. The different interpretations of the Jiang *et al.* paper by these groups indicate the importance of prior beliefs and interests in shaping which knowledge claims will be accepted by whom.

A second way in which stem cells are classified is according to their therapeutic potential only, not considering their differentiation potential: some scientists argue that the best stem cells are the ones that will lead to cures or at least therapies, regardless of how. For instance, another Australian researcher, referring to the controversy over the mechanism explaining the results published by Jiang *et al.* (2002) argues:

> Rachel: There's been a lot of controversy in the stem cell field. Some people just do not believe that blood, bone marrow stem cells, can actually form other tissues. Other people believe it's true. We certainly find evidence that at a very low percentage, cells from the kidney are coming in and forming [cells from another tissue]. So I think it happens, but not very often. So whether this is trans-differentiation, [. . .] [or] whether it's fusion for example [. . .]. Like, it might just be part of the repair and remodelling process that tissue goes through during inflammation. But that's fine [. . .].

For Rachel, what matters is not whether AS cells are pluripotent and can 'trans-differentiate'. Rather, it is that they can contribute to injury repair and thus have clinical applications – this is the area she is working in. Useful stem cells for her can contribute to organ repair, and do not need to have the highest differentiation potential. In contrast to scientists above, who challenged the materiality of the pluripotent AS cell but maintained the importance of pluripotency, Rachel argues that differentiation potential is not essential for therapeutic potential. Since the latter is the only important criterion in her classificatory scheme, the former becomes irrelevant.

Bernard, a firm supporter of ASCR, also focuses on therapies: he conducts much of his boundary-work to 'monopolize' resources (eg Gieryn, 1983: 788–9) for ASCR by focussing on its therapeutic applications, and dismisses the clinical potential of ESCR as hype:

> Bernard: the debate was being driven by the promises of using [ESCR] for therapy, which seemed to be a little premature at that stage, I mean the promise is always there, but the promise is there for anything you can think about, and the idea that you want to cure disease, doesn't translate into the fact that you can do it.

He is critiquing the use of 'promise' in the same way that Kitzinger and Williams (2005) critique the use of 'hope' in discourses promoting embryo research: proponents of controversial research conjure images of 'limitless and imminent potential', without having to provide any hard evidence. Hence, if nothing materializes, they have an 'escape clause' as they were only ever talking about possibilities, not making forecasts (2005: 738). This emphasizes the central role of 'expectations of future breakthroughs' in mobilizing support for particular areas of innovative research (as Brown and Kraft, 2006, found in the context of cord blood banking). When past 'promises' are re-examined by others, particularly from a competing field, their over-use can be criticized, especially if it leads to 'disappointment' (Brown and Michael, 2003: 12). This is what Bernard does here. For him, and others belonging to the same 'community of promise', the promissory stories are merely strategic and the 'therapeutic ES cell' lacks ontological robustness.

Thirdly, some scientists classify stem cells according to their differentiation potential and their stability *in vivo*: they argue that the best stem cells are the ones that have a *lower* and, more importantly, controllable differentiation potential. Indeed, lack of control can lead to cancers when these cells are implanted into patients. The pluripotency of ES cells was used as a negative point by opponents of ESCR during Australian parliamentary debates and interviews. For instance, a junior Australian researcher who works on AS cells, states:

> Nathan: I think adult cells are miles ahead, miles, miles, miles ahead. I mean people have had [. . .] bone marrow grafts for years, all of the animal work shows that they don't make tumours, there's no problem, it's a benign thing in a way. Whereas [. . .] I've seen a rat, they put [ES cells] in the brain or something, and it's growing right out of his head, walking around with this big growth sitting out of its skull. That's an extreme case, but that's sort of how it is now.

Nathan excludes ES cells from his category of most promising stem cells and tries to promote the authority of AS cells, which he describes as safer.

In the UK too, the importance of control was discussed during interviews. For example:

> Ted: [. . .] we know [ES cells have] got a potentiality, in some ways they've got too much potentiality. The problem with adult stem cells is, we know that they're not carcinogenic but do they [have] a potentiality? [. . .] And I suspect at the end of the day, it's not going to be either adult stem cells or human ES cells, it's going to be some kind of mix of the two or one informed by the other.

Ted suggests that finding pluripotent cells is not the quintessential goal of SCR. In contrast to Nathan, however, he does not use the risk of cancer as a reason to monopolize resources for ASCR and marginalize ESCR. Rather he promotes SCR in general: all these problems can be overcome by further research (see also Parry, 2009) thus expanding the authority of SCR as a whole (cf Gieryn, 1983: 783–7, 1999: 37–64).

Interestingly, although this issue of safety and control came up during these interviews and has been discussed in the media (eg Janzen and Scadden, 2006), it did not feature prominently in public debates before 2001. The contrast between a UK-based scientist, Ted, mentioning this issue here, and the absence of it in story A (which, as indicated above, dominated UK public discussions before 2001) can be explained in two ways. Firstly, it is easier to talk about risks in the relatively safe environment of the private interview. Secondly, the public debates and this interview took place at different points in time: the former took place when the future of ESCR was highly contested, the latter when it was more settled. This second explanation is reinforced by evidence that Ted did focus on the issue of control during a public talk he gave later in 2005. This highlights the temporality of the dynamics of expectation where 'radical discourses' are more important when a technology is seen to be emerging and its future unsettled (Brown and Michael, 2003: 16). The example of classifying cells according to differentiation potential and control also shows that an acceptance of the ontological reality of uncontrolled ES cell tumours does not automatically lead to a commitment to AS cells over ES cells.

A fourth classification criterion draws on the concept of 'nature'. This can be a powerful rhetorical tool. For example, Parry (2003a) shows how publics draw on their lived experiences to construct the category of 'nature' which they can use in contingent ways in their discourses on SCR and cloning. Brown and Michael (2003) discuss how professionals use 'nature': they explain the technical difficulties encountered by xeno-transplantation and its limited success by pointing to its 'unnaturalness'; they contrast this to SCR's predicted success, due in part to its 'naturalness'. Here, particular areas of research that people support and are comfortable with are portrayed as natural, while research considered unacceptable is portrayed as unnatural.

Nature was used to 'monopolize' resources for SCR by excluding reproductive cloning as 'unnatural', in contrast to a more 'natural' reproductive method: IVF. One researcher states in an interview that he does not find reproductive cloning 'acceptable', because 'it just doesn't feel right to me'. When asked if it would be acceptable in the context of treating infertility, he suggests that infertility should only be treated by what he calls 'natural' means such as IVF and adoption. He does not reflect on processes of 'naturalization' that technologies such as IVF have gone through (cf. Parry, 2003a, see also Brown, 1999 on 'normalization'), but rather posits 'nature' as a stable category. He uses the idea of nature as a classification criterion: what is seen as 'natural' maps onto what he describes as acceptable, whereas what is 'unnatural', including reproductive cloning, is not acceptable.

Invoking 'nature' is also an authoritative way of doing adult/embryonic boundary-work. It can be used to promote ASCR at the expense of ESCR:

> Zach: The sort of idea that the embryonic are much more . . . pluripotent, and have [a] much greater range of differentiation, but, I don't know how that relates . . . I think, I like the idea of, with my work, that we're not trying to push them too far anyway. Like we're trying to encourage them to go down their *natural* pathway, just more efficiently I suppose. (emphasis added)

Zach is a PhD student working on a particular type of AS cell, trying to understand their differentiation in more detail. He suggests that it is more 'natural' to work with AS cells than trying to 'push' ES cells down particular differentiation pathways that they would not follow without human intervention. In his account, by painting it as an attempt to make cells do something they would not 'naturally' do, ESCR is marginalized. Accounts such as these illustrate one discursive process through which cognitive authority is secured in one area of SCR (in this case, ASCR) by invoking particular ideas of 'nature' and the 'natural'.

In contrast, an ES cell scientist in the UK drew the map of good SCR to include ESCR and expel ASCR. He did not explicitly appeal to the label 'nature', but talked about what cells are 'meant' to do. For him, ES cells are 'meant' to 'generate all the cell types' in the body. Conversely, AS cells are not, since there is no 'physiological function' for them to do so; they do not do this during the normal course of development. According to this scientist, researchers should not try to force AS cells to give rise to all cell types in the body, and the idea of pluripotent AS cells is labelled 'hype'. By talking about what cells are 'meant' to do, he appeals to a higher order, perhaps nature, that guides which research is appropriate.

Thinking in terms of goals and interests, it would be easy to end this section by arguing that classification criteria are always chosen instrumentally to advance one's sub-field of research. However, the situation is more complex. For example, some researchers who work on AS cells refuse to argue that a high differentiation potential for those cells is enough to stop work on ES cells, instead pushing for research on all cells. This line of argument is often taken to 'protect the autonomy' of science (Gieryn, 1983: 789–91) and exclude ethical reasons from becoming relevant to the choice of future research directions. Also, the choice of selection criteria may not reflect narrow professional interests, but rather highlight broader views about the role of science: an interest in differentiation potential indicates a focus on using cells as research tools, whereas an interest in better control indicates a focus on using cells for therapies.

Discussion and conclusions

When definitions or stories are told by authoritative figures in public arenas, some can stabilize and seem to be accurate reflections of reality – given by nature

– rather than social constructions. The present chapter challenges this. By capturing some of these stories in the making and by focussing on the multiple, messy and socially located ways in which they come about and are accepted within particular communities, it unveils the ontological fluidity of nature, maps out some of the interests of those telling stories and draws out some of the material implications of the stabilization of particular classifications of nature. It also shows how the concepts of nature or 'natural' can be drawn on to give authority to particular classifications.

Using the example of stem cells, and drawing in particular on Barnes' approach to classifications, the contingency of nature is explored and accounts of the indeterminacy of 'ontological ordering' (Jasanoff, 2005) and labelling (eg Mulkay, 1994; Hauskeller, 2005b; Parry, 2009) are further developed. In addition, Gieryn's conceptual tool of 'cultural cartography' (Gieryn, 1999) is used to examine which boundaries are erected around and within SCR, and how these rhetorical constructions can be drawn on in attempts promote the authority of particular subfields within SCR, or of the field as a whole. Thus boundary-work is a valuable way of exploring not only science as a whole, but also particular areas within science.

The under-determinacy of 'stemness' enables different conceptualizations of SCR to be expressed. One way of talking about SCR is put forward, story A; it was dominant during British public debates in 2001. This story is contrasted with a number of alternatives which took hold in other places, at other times, within other communities. By paying attention to actors' social locations and relating them to their stem cell stories, the ways in which particular classifications may better map onto particular interests is explored, therefore suggesting reasons for the acceptance of these stories in particular communities. Specifically, the temporality of stem cell stories and of the dynamics of expectations is illustrated.

Four main classification criteria are drawn upon by stem cell researchers here. 'Differentiation potential', which is under-determined by reality, is one. 'High differentiation potential' can indicate high therapeutic and research value for both ESCR and ASCR, but also low therapeutic and research value for both ESCR and ASCR. Whether a group of stem cells is considered to be promising within this classification scheme depends on the belief in the 'ontological robustness' of particular entities, such as the 'pluripotent AS cell'. So, when using differentiation potential as a criterion, if one accepts the existence of the pluripotent AS cell, one sees AS cells as the best cells to study, but if one rejects its existence, one sees ES cells as the best cells.

For other researchers, the materiality of pluripotent AS cells is irrelevant, and only 'therapeutic potential' matters. This criterion, with a similar fluidity to 'differentiation potential', can be used to promote or demote both adult and embryonic SCR. For yet other scientists, safety and control are the essential criteria by which to judge stem cells. These are also flexibly applied to promote the sub-field of ASCR as well as SCR as a whole.

In addition, 'nature' is a useful interpretive resource that promotes different areas of research (IVF, IVF-derived ESCR, ASCR) and exclude others (reproductive cloning, ASCR, ESCR). The constructedness of 'nature' is visible here; however, none of my informants reflect upon it, unlike lay groups examined by Parry (2003a; see also Milne, this volume).

Scientists are also shown to conduct boundary-work to 'protect their autonomy' from outside influences and are not slaves to their narrow professional interests. Often, maps drawn do not directly reflect a field of research; rather, they defend the authority of science or of SCR as a whole

The present account has attempted to eschew both purely 'normative' and 'descriptive' approaches to scientific knowledge (Fagan, 2007a): scientists' definitions and knowledge claims are not seen as mere opinions, or as absolute truths existing outside society. Rather, epistemic success is explained in the context of social goals and interests (see also Fagan, 2007b). Of particular importance, are 'communities of promise' and the role of expectations, in explaining how stories stabilize, at least temporarily. Some groups of people share visions of the present and future. Within these communities, particular boundaries and classification criteria take on importance, and particular entities are seen as 'ontologically robust' (Jasanoff, 2005). For instance, a group that shares an imagined future where research on adult stem cells keeps Australia at the forefront of biotechnological research might more readily accept the ontological robustness of the pluripotent AS cell. Its existence might be challenged and labelled as hype by those who have a contradictory vision of the future, perhaps based on embryo research.

Having highlighted the diversity of ways to describe stem cells and SCR, it now becomes important to make some normative points. As Castree and Braun argue, Sociology of Scientific Knowledge scholars can show 'where responsibility lies, what stories need to be told and what differences these make' (1998: 33). In the context of calls for public engagement in science and technology (eg House of Lords, 2000; Jasanoff, 2003; Leach *et al.*, 2005; Irwin, 2006; Joly and Kaufmann, 2008; Davies *et al.*, 2009), it is essential to think about the implications of these multiple stories for science-public relations. The chapter has shown the 'facts' of SCR to be fluid and contingent. Therefore, so-called public 'ignorance' of these science 'facts' partly reflects these multiple possible stories and should not be a barrier to public engagement in technoscience. Consequently, it is essential to continue challenging the sometimes dominant framings of public engagement events by science.

In addition, many classifications focus on stem cell attributes that scientists are expert in finding and defining (not everyone can attempt to measure a cell's differentiation potential). This can increase their cognitive authority and displaces conversations about other criteria such as the source of the stem cells, ethics, social pressure to donate body parts, cancer risks etc. To paraphrase Star (1985): the more stem cell researchers argue with one another about *how* to use stem cells (such as adult or embryonic), the less salient the question of *whether* to use them becomes.[10] So by focussing on whether adult cells are 'pluripotent' or merely 'multipotent', we are not asking if money could be better spent on recruiting carers, improving access for wheelchair users or promoting healthy eating, for example. This needs to be further examined.

Acknowledgements

This research was funded by an MRC studentship. Support from the MRC Human Genetics Unit and the ESRC Innogen Centre is gratefully acknowledged. Particular thanks go to Sarah Parry for supervising me throughout my PhD and helping with the present manuscript. I would also like to thank Wendy Faulkner and Veronica Van Heyningen for supportive supervision during my PhD. Thanks go to Jane Calvert, Barry Barnes, John Dupré and Christine Hauskeller and anonymous reviewers for insightful and constructive comments on earlier drafts. Finally, thank you to all the scientists who accepted to be part of this research. All mistakes and inconsistencies are my own.

Notes

1 It is beyond the remit of this chapter to explore all the potential classification schemes and definitions used. For instance, the differences between 'cloned' and '*left-over* IVF embryos', between embryos and foetuses, or between 'multipotent' and 'pluripotent' stem cells are not detailed here. These are discussed elsewhere (Marks, 2008; see also Hauskeller, 2005b).

2 See www.ISSCR.org for further details of current legislation.

3 As suggested by Parry (2003a: 198), many people held more nuanced positions than simply for or against ESCR. Nevertheless, these debates took place in the lead up to votes for or against research on embryos.

4 The legitimacy of this work is still at stake in other public *fora* however.

5 See the Research Involving Human Embryos Act (Commonwealth of Australia, 2002a) and the Prohibition of Human Cloning Act (Commonwealth of Australia, 2002b). For an analysis, see Morley and Hall, 2003; Harvey, 2005; Robins, 2005; Dodds and Ankeny, 2006. The ban on human embryo cloning (or therapeutic cloning, or somatic cell nuclear transfer) was overturned after the end of data collection (Commonwealth of Australia, 2006). See Harvey, 2008; Ankeny and Dodds, 2008, for further information.

6 This was a controversial claim as it challenged one of the fundamental principles of biology, that once cells specialize into a particular cell type, they lose the ability to turn into other cells types. It was previously thought that once a cell committed to being a blood stem cell, for example, it could then only give rise to different kinds of blood cells, not skin or nerve cells.

7 AS cells were also said to have a higher degree of 'plasticity' than previously thought. The labels 'plasticity', 'high differentiation potential' and 'pluripotency' are often used interchangeably. There is also a certain amount of slippage between the labels 'pluripotent' and 'multipotent'; for instance, Jiang *et al.* (2002) describe these stem cells as having 'pluripotency' in the title, but as being 'multipotent' in the abstract (see Marks, 2008: 92–5 for further information).

8 All the scientists interviewed are given pseudonyms (as in Gilbert and Mulkay, 1984). Although some could be identified by people familiar with SCR, pseudonyms are used for two reasons. Firstly, in accordance with the informed consent discussions held with participants, the aim is to protect their identity as far as possible. Secondly, the focus of this chapter is on *general* pro-

cesses of defining and clustering, not those of particular protagonists. The use of first name pseudonyms rather than surnames also has two reasons. Firstly, in the context of a study committed to the importance of alternative forms of knowledge and public engagement, it was important not to reify the dominance of scientific expertise by using titles. Secondly, the author has spent a lot of time in Australia where the use of first names is common and not felt to be unduly familiar.

9 The findings from Jiang *et al.* (2002) continue to be disputed. However, a new but similar challenge has risen against ESCR in the form of 'induced pluripotent stem cells' which come from adult tissues and can relatively easily be given a similar differentiation potential to ES cells under laboratory conditions (Aoi *et al.*, 2008; Ebert *et al.*, 2009).

10 The original is: 'The more localizationists argued with one another about *how* to do, for example, ablation experiments, the less salient the question of *whether* to do them became' (Star, 1985: 412).

References

Adams, G.B. and D.T. Scadden, (2006), 'The Hematopoietic Stem Cell in Its Place', *Nature Immunology*, 7: 333–7.

Alison, M., R. Poulsom, W. Otto, P. Vig, M. Brittan, N. Direkze, S. Preston and N. Wright, (2003), 'Plastic Adult Stem Cells: Will they Graduate from the School of Hard Knocks?', *Journal of Cell Science*, 116: 599–603.

Ankeny, R. and S. Dodds, (2008), 'Hearing Community Voices: Public Engagement in Australian Human Embryo Research Policy, 2005–2007', *New Genetics and Society*, 27: 217–32.

Aoi, T., K. Yae, M. Nakagawa, T. Ichisaka, K. Okita, K. Takahashi, T. Chiba and S. Yamanaka, (2008), 'Generation of Pluripotent Stem Cells from Adult Mouse Liver and Stomach Cells', *Science*, 321: 699–702.

Barnes, B., (1983), 'On the Conventional Character of Knowledge and Cognition,' in Knorr-Cetina, K.D. and M. Mulkay (eds.), *Science Observed: Perspectives on the Social Study of Science*, London: Sage.

Barnes, B., (2002), 'Searle on Social Reality: Process Is Prior to Product', in Grewendorf, G. and G. Meggle (eds), *Speech Acts, Mind, and Social Reality: Discussions with John R. Searle*, Dordrecht: Kluwer Academic Publisher.

Barnes, B., D. Bloor and J. Henry, (1996), *Scientific Knowledge: A Sociological Analysis*, London: Athlone.

Barnes, B. and D. Edge, (eds), (1982), *Science in Context: Readings in the Sociology of Science*, Milton Keynes: Open University Press.

Bender, W., C. Hauskeller and A. Manzei, (eds.), *Crossing Borders: Cultural, Religious and Political Differences Concerning Stem Cell Research, A Global Approach*, Münster: Agenda Verlag.

Bloor, D., (1976), *Knowledge and Social Imagery,* 2nd edn, London: Routledge and Kegan Paul.

Borup, M., N. Brown, K. Konrad and H.V. Lente, (2006), 'The Sociology of Expectations in Science and Technology', *Technology Analysis & Strategic Management*, 18: 285–98.

Brown, N., (1999), 'Xenotransplantation: Normalizing Disgust', *Science as Culture*, 8: 327–57.

Brown, N. and A. Kraft, (2006), 'Blood Ties: Banking the Stem Cell Promise', *Technology Analysis & Strategic Management*, 18: 313–27.

Brown, N. and M. Michael, (2003), 'A Sociology of Expectations: Retrospecting Prospects and Prospecting Retrospects', *Technology Analysis and Strategic Management*, 15: 3–18.

Castree, N. and B. Braun, (1998), Chapter 1: The Construction of Nature and the Nature of Construction: Analytical and Political Tools for Building Survivable Futures. *Remaking Reality: Nature at the Millenium*, London: Routledge.

Collins, H.M. and T.J. Pinch, (1993), *The Golem: What Everyone Should Know About Science*, Cambridge: Cambridge University Press.

Commonwealth of Australia, (2002a), *Research Involving Human Embryos Act*, Canberra: The Parliament of the Commonwealth of Australia.

Commonwealth of Australia, (2002b), *Prohibition of Human Cloning Act*, Canberra: The Parliament of the Commonwealth of Australia.

Commonwealth of Australia, (2006), *Prohibition of Human Cloning for Reproduction and the Regulation of Human Embryo Research Amendment Act*, Canberra: The Parliament of the Commonwealth of Australia.

Davies, S., E. McCallie, E. Simonsson, J.L. Lehr and S. Duensing, (2009), 'Discussing Dialogue: Perspectives on the Value of Science Dialogue Events that do not Inform Policy', *Public Understanding of Science*, 18: 338–53.

Dodds, S. and R.A. Ankeny, (2006), 'Regulation of Hesc Research in Australia: Promises and Pitfalls for Deliberative Democratic Approaches', *Journal of Bioethical Inquiry*, 3: 95–107.

Ebert, A.D., J. Yu, F.F. Rose, V.B. Mattis, C.L. Lorson, J.A. Thomson and C.N. Svendsen, (2009), 'Induced Pluripotent Stem Cells from a Spinal Muscular Atrophy Patient', *Nature*, 457: 277–80.

Eriksson, L. and A. Webster, (2008), 'Standardizing the Unknown: Practicable Pluripotency as Doable Futures', *Science as Culture*, 17: 57–69.

Fagan, M.B., (2007a), 'The Search for the Hematopoietic Stem Cell: Social Interaction and Epistemic Success in Immunology', *Studies in History and Philosophy of Science Part C: Studies in History and Philosophy of Biological and Biomedical Sciences*, 38: 217–37.

Fagan, M.B., (2007b), Social Epistemology of Scientific Inquiry: Beyond Historical Vs. Philosophical Case Studies. http://philsci-archive.pitt.edu/archive/00003586/01/mbf_&HPS1.pdf (accessed 13 Jan 10).

Gieryn, T., (1983), 'Boundary-Work and the Demarcation of Science from Non-Science: Strains and Interests in Professional Ideologies of Scientists', *American Sociological Review*, 48: 781–95.

Gieryn, T., (1995), 'Boundaries of Science', in Jasanoff, S., G. Markle, J. Petersen and T. Pinch (eds), *Handbook of Science and Technology Studies*, London: Sage.

Gieryn, T., (1999), *Cultural Boundaries of Science: Credibility on the Line*, Chicago, London: Chicago University Press.

Gilbert, N.G. and M. Mulkay, (1984), *Opening Pandora's Box: A Sociological Analysis of Scientists' Discourse*, London: Cambridge University Press.

Goven, J., (2006), 'Processes of Inclusion, Cultures of Calculation, Structures of Power; Scientific Citizenship and the Royal Commission on Genetic Modification', *Science, Technology, & Human Values*, 31: 565–98.

Haraway, D.J., (1997), *Modest_Witness@Second_Millenium.Femaleman©Meets_Oncomouse™: Feminism and Technoscience*, Routledge: New York.

Harvey, O., (2005), 'Regulating Stem-Cell Research and Human Cloning in an Australian Context: An Exercise in Protecting the Status of the Human Subject', *New Genetics & Society*, 24: 125–36.

Harvey, O., (2008), 'Regulating Stem Cell Research and Human Cloning in an Australian Context: The Lockhart Review', *New Genetics and Society*, 27: 33–42.

Hauskeller, C., (2005a), 'The Language of Stem Cell Science', in Bender, W., C. Hauskeller and A. Manzei (eds), *Crossing Borders: Cultural Religious and Political Differences Concerning Stem Cell Research, A Global Approach.* Münster: Agenda Verlag.

Hauskeller, C., (2005b), 'Science in Touch: Functions of Biomedical Terminology', *Biology and Philosophy*, 20: 815–35.

Hauskeller, C., W. Bender and A. Manzei (eds), (2005), *Crossing Borders. Grenzüberschreitungen*, Munster: Agenda Verlag.

House of Lords, (2000), Science and Technology Committee, Third Report, *Science and Society*, Session 1999–2000, HL38.

Human Fertilisation and Embryology Act, (1990), Statutory Instrument 1991, No. 1781.

Human Fertilisation and Embryology (Research Purposes) Regulations, (2001), Statutory Instrument 2001, No. 188.

Irwin, A., (2006), 'The Politics of Talk: Coming to Terms with the "New" Scientific Governance', *Social Studies of Science*, 36: 299–320.

Janzen, V. and D.T. Scadden, (2006), 'Stem Cells: Good, Bad and Reformable', *Nature* 441(7092): 418–19.

Jasanoff, S., (2003), Technologies of Humility: Citizen Participation in Governing Science. *Minerva*, 41: 233–44.

Jasanoff, S., (2005), *Designs on Nature: Science and Democracy in Europe and the United States*, Princeton, NJ: Princeton University Press.

Jiang, Y., B.N. Jahagirdar, R.L. Reinhardt, R.E. Schwartz, C.D. Keene, X.R. Ortiz-Gonzalez, M. Reyes, T. Lenvik, T. Lund, M. Blackstad, J. Du, S. Aldrich, A. Lisberg, W.C. Low, D.A. Largaespada and C.M. Verfaillie, (2002), 'Pluripotency of Mesenchymal Stem Cells Derived from Adult Marrow', *Nature*, 418: 41–9.

Joly, P.B. and A. Kaufmann, (2008), 'Lost in Translation? The Need for "Upstream Engagement" with Nanotechnology on Trial', *Science as Culture*, 17: 225–47.

Kitzinger, J. and C. Williams, (2005), 'Forecasting Science Futures: Legitimising Hope and Calming Fears in the Embryo Stem Cell Debate', *Social Science & Medicine*, 61: 731–40.

Lajtha, L.G., (1983), 'Stem Cell Concepts', in Potten, C.S. (ed.), *Stem Cells: Their Identification and Characterisation*, Edinburgh: Churchill Livingstone.

Leach, M., I. Scoones and B. Wynne, (eds.), (2005), *Science and Citizens: Globalization and the Challenge of Engagement*, London: Zed Books.

Marks, N., (2008), *Opening up Spaces for Reflexivity? Scientists' Discourses About Stem Cell Research and Public Engagement*, University of Edinburgh, unpublished PhD thesis.

Martin, P., N. Brown and A. Kraft, (2008), 'From Bedside to Bench? Communities of Promise, Translational Research and the Making of Blood Stem Cells', *Science as Culture*, 17: 29–41.

Morley, K.I. and W. Hall, (2003), 'Regulation of Embryonic Stem Cell Research and Therapeutic Cloning: The Australian Debate', *Plaintiff*, 55: 20–23.

Mulkay, M., (1994), 'Triumph of the Pre-Embryo: Interpretations of the Human Embryo in Parliamentary Debate over Embryo Research', *Social Studies of Science*, 24: 611–39.

Nelkin, D., (1975), 'The Political Impact of Technical Expertise', *Social Studies of Science*, 5: 35–54

Parliament of Australia, S., (2002), Provisions of the Research Involving Embryos and Prohibition of Human Cloning Bill; Supplementary Report in Favour of the Legislation, http://www.aph.gov.au/Senate/committee/clac_ctte/completed_inquiries/2002–04/emb_cloning/report/d02.htm (accessed 13 Jan 10).

Parry, S., (2003a), *Debating Stem Cell Research and Human Cloning: Multiple Meanings, Competing Constructions*, University of Edinburgh, unpublished PhD thesis.

Parry, S., (2003b), 'The Politics of Cloning: Mapping the Rhetorical Convergence of Embryos and Stem Cells in Parliamentary Debates', *New Genetics and Society*, 22: 145–68.

Parry, S., (2009), 'Stem Cell Scientists' Discursive Strategies for Cognitive Authority', *Science as Culture*, 18: 89–114.

Robins, R. (2005), 'Biomedical Innovation or Bioethical Precaution: The Stem Cell Debate in Australia', in Bender, W., C. Hauskeller and A. Manzei, (eds.), *Crossing Borders: Cultural, Religious and Political Differences Concerning Stem Cell Research, A Global Approach*, Münster: Agenda Verlag.

Rubin, B.P., (2008), 'Therapeutic Promise in the Discourse of Human Embryonic Stem Cell Research', *Science as Culture*, 17: 13–27.

Science as Culture, (2008), 'Special Issue: Stem Cell Stories 1998–2008', *Science as Culture*, 17.

Selin, C., (2007), 'Expectations and the Emergence of Nanotechnology', *Science, Technology, & Human Values*, 32: 196–220.

Shapin, S., (1995), 'Trust, Honesty, and the Authority of Science', in Bulger, R.E., E.M. Bobbya and H. Fineberg (eds), *Society's Choices: Social and Ethical Decision Making in Biomedicine*, Washington DC: National Academy Press.

Star, S.L., (1985), 'Scientific Work and Uncertainty', *Social Studies of Science*, 15: 391–427.

Trounson, A., (2006), 'The Production and Directed Differentiation of Human Embryonic Stem Cells', *Endocrinological Reviews*, 27: 208–19.

Van Dyck, J., (1995), *Manufacturing Babies and Public Consent: Debating the New Reproductive Technologies*, Basingstoke: MacMillan.

Wynne, B., (1996), 'May the Sheep Safely Graze? A Reflexive View of the Expert-Lay Knowledge Divide', in Lash, S., B. Szerszynski and B. Wynne (eds), *Risk, Environment and Modernity: Towards a New Ecology*, London: Sage.

Wynne, B., (2006), 'Public Engagement as a Means of Restoring Public Trust in Science - Hitting the Notes, but Missing the Music?' *Community Genetics*, 9: 211–20.

Zipori, D., (2005), 'The Stem State: Plasticity Is Essential, Whereas Self-Renewal and Hierarchy Are Optional', *Stem Cells*, 23: 719–26.

Part Three
(Re)Modelling Nature

Captivating behaviour: mouse models, experimental genetics and reductionist returns in the neurosciences

Gail Davies

Introduction

In December 2007, the behaviour of a genetically altered mouse, apparently expressing no fear of cats, was demonstrated in the media. The fearless mouse, generated at the laboratories of Professor Kobayakawa at the University of Tokyo, was created through a genetic mutation which shut down receptors in the mouse olfactory bulb – the part of the brain that processes information about smell – so the animal no longer reacted to the scent of its natural predator. The press appeal of the story was amplified by the release of a short video, posted widely on the internet.[1] The video, shot in a nondescript room, shows a black mouse nonchalantly approaching a tabby cat, walking underneath the larger animal and nuzzling it; it is apparently undaunted as it moves rapidly and repeatedly between the cat's front paws. The mouse's creator describes the experiment, and its meaning for him, in the accompanying *Guardian* article: '"The mice approached the cat, even snuggled up to it and played with it", Kobayakawa said. "The discovery that fear is genetically determined and not learned after birth is very interesting, and goes against what was previously thought"' (McCurry, 2007). We are told his conclusions from this experiment: confirming suspicions about the relationship between scent and fear, and suggesting the natural tendency for mice to fear cats is genetically determined. Kobayakawa's hopes for the application of this research in neuroscience are reported in *The Telegraph*: 'We think it has the power to clarify many unrevealed principles of the brain, those which generate emotions and behaviours in mammals' (Highfield, 2007); mammals, of course, including humans.

This is just one example of the well-rehearsed scientific press release in which a genetically altered mouse demonstrates the potential genetic components of a relevant human behaviour. Since the 1920s, the mouse has played a critical role in understanding animal and human genetics (Rader, 2004; de Chadarevian, 2006). This prominence has grown with the development of transgenic mice from the 1980s, the on-going focus on genetics in understanding human disease, the mapping of mouse and human genomes and the expansion of national and regional centres for archiving and distributing genetically altered mice (Grimm, 2006; Shostak, 2007). Since the late 1990s, the growing availability of genetically altered strains, developed to exhibit certain behaviours, means the genetically altered mouse is taking over from the more assertive and archetypal 'lab rat' in the search for better models of human behaviours.

Many of these developments are similarly reported in the media. Recent news stories reveal mice who are abnormally aggressive, which might be used to understand criminality; mice portraying symptoms of schizophrenia, to understand and treat this mental illness; mice prone to overeating, to treat obesity, and mice who fidget, to understand weight maintenance; there are mice who are depressed and mice who are anxious, which may help understand this spectrum of the human condition; and there are forgetful mice, which may yield insights into Alzheimer's disease. Some animals have gained popular names as well as media coverage: the so-called 'mighty mouse' or supermouse, who was bred to aid understanding of the PEPCK-C enzyme, but whose ability to run many miles without tiring, eat substantial amounts without weight gain, and have more sex and at older ages was greeted with admiration and some envy by media commentators. The announcement of such animals suggests that the ability of scientists to understand and manipulate natural behaviour is being radically transformed. Yet these media reports of scientific experimentation are yet to be followed by news of clinical application. In fact, they often lack substantive detail. The complex interactions between animal strain, physical environment, genetic modification, experimental design, drug intervention and laboratory handling that influence behaviour are not introduced.

These complexities can be illustrated by going back to the video of the fearless mouse. The relationship between genetics and behaviour is open to alternative interpretations if attention shifts from the laboratory mouse to the bemused, but submissive cat, which patiently shifts position as the mouse moves around its body. The cat is not genetically altered. Rather the article suggests the video features Mochikko-chan, a domestic pet, belonging to one of the research team. The cat was chosen as especially docile and unlikely to end this uncharacteristic encounter between species. With this additional information, the meaning of the video becomes more ambiguous. The mouse has no need to fear this unusually compliant cat: perhaps this cat's behaviour is the less natural. The causality of animal behaviour is opened up from this focus on genetics. The genetic modification of the mouse is foregrounded, but the reason the cat is so quiescent is not explained in the video. We accept that pets can be handled, neutered, well-fed and otherwise habituated so they do not readily express what might be seen as natural hunting instincts. Yet corresponding questions about whether laboratory mice may be similarly accustomed through breeding, husbandry and conditioning to express particular behaviours, and indeed what is natural behaviour for a laboratory animal, are not addressed.

These questions begin to take on an importance, which exceeds analysis of this short video, when they are put into a wider context. Behavioural neuroscience is one of the fastest growing and potentially lucrative areas of biomedical research. There are many who have the potential to gain from innovative treatments for anxiety, depression and other affective disorders, which are some of the most widespread health problems in the West. Yet, as researchers themselves suggest 'psychiatry has proven to be among the least penetrable clinical disciplines for the development of satisfactory *in vivo* model systems for evaluating novel treatment approaches' (Cryan and Holmes, 2005: 775). Nevertheless, the development of genetically altered animal models of affective disorders remains a focus for much of this research. The market for such drugs is expanding rapidly as the West ages and the mental health needs in the rest of the world become more like those of the West. The potential rewards are also high as treatment for these conditions is on-going, rather than one-off. Thus, these media reports reflect not only scientific research but also contemporary social concerns and promissory pharmaceutical futures. The presentation and circulation of these genetically altered animals become part of what Dumit (2004) calls the production of 'surplus health'; the turning of everyone into patients or consumers-in-waiting, who can be put on drugs for life (cited in Sunder Rajan, 2006; see also Lakoff, 2005). It is not only the nature of animal behaviour being transformed, but also our understandings of human nature and our definitions of what is normal and what is treatable.

This chapter looks at the relationship between human and animal behaviour, as they are enmeshed in and emerge from the contemporary behavioural neurosciences. This is an area of both scientific investigation and philosophical speculation. It raises questions about the nature of behaviour, which I introduce in the next section, exploring how media framing of these scientific discourses tends to reinscribe older dualities between nature and nurture. It also raises questions about how to challenge these dualities, which I explore theoretically, through work drawing on early ethology as a prelude to more relational accounts of animals and their environment, and empirically, through tracing the complex interplay of animal and human agencies within the experimental spaces of the behavioural neurosciences. Through reference to the work of Uexküll and others, social scientists have sought new ways of addressing questions around animal and human behaviour. This provocation has been of considerable interest in anthropology (Ingold, 2000), geography (Thrift, 2005) and cultural studies (Broglio, 2008), leading to suggestions of an ethological turn in social theory (Lorimer, 2008). This is not a singular theoretical encounter and it is sometimes a contradictory impulse, at times both challenging humanism and confirming anthropocentrism. Yet the work of Uexküll, as taken up by Heidegger and Agamben, has become a key engagement for the thinking about the animal question (Elden, 2006). Here I take Agamben's 2004 work *The Open* – an ontological genealogy of the articulations and divisions between humans and animals – as provocation of where to look in tracing the transaction of capacities between animals and humans in behavioural genetics. The empirical account which follows traces divergent meanings of mouse behaviour as they emerge from the practices of scientists studying in different spaces, laboratories and paradigms, reflecting on the points of connection and disruption between them.

In so doing, it introduces the role of these genetically altered animals in stabilizing relations around particular trajectories of drug development, whilst also offering points at which it might be possible to think, or behave, otherwise. Concluding, I return to the philosophical analysis offered by Agamben, finding his attention to the ontologies and topologies of the biosciences productive, but seeing little that adds to our understanding of animal, as opposed to human, nature. Instead, preliminary observations from the boundaries of ethology and anthropology suggest how we might add forms of life and understandings of nature to the experimental practices of the behavioural neurosciences.

Nature, behaviour and captivation

I start with reflections on the concept of nature in the context of behavioural genetics and the neurosciences. Nature retains a powerful place within political debates about biotechnology (Davies, 2006; Hansen, 2006), with concepts of nature conversely used to encapsulate what is morally troubling about new developments or designate what is merely an extension of existing practices in the genetic modification of plants, animals and humans. Nature is mobilized in the contexts of health, as both fact and value, in the classification of natural and abnormal states (Canguilhem, 1991). Further, behavioural genetics returns us to freighted and enduring arguments about the contributions of nature and nurture to the development of particular patterns of behaviour. The continued use of arguments about nature and nurture engenders frustration in scientists. As Keller remarks, the debate is felt to be long resolved: 'everything is a mixture of both, and it is all very complicated. Why then do we not save our energies for more legitimate, and more fruitful, areas of scientific inquiry?' (Keller, 2008: 118). It is this query I want to address first. For, as Keller suggests, ideas about nature as separate from nurture endure in the models, metaphors, drugs and devices used in experimental practices and scientific claims, particularly in the study of behavioural genetics.

To elaborate, ideas about nature appear in at least three distinct ways in the above encounter between the cat and the genetically altered mouse; broadly mapping onto Raymond Williams' three-fold classification of ideas of nature as 1) the essence of something, 2) nature as universal – relating to the physical world in general – or 3) as that which is unaltered by human intervention (Williams, 1983). First, there is the idea of nature as essence, character or kind. The meaning of the video depends on our understanding of the specific nature of different species, minimally of mouse as prey and cat as predator. This nature is given by the animal's *telos*, or alternatively its evolutionary history. It is based in biology, but it is also plural and differentiated, such that the nature of a mouse, the nature of a cat and the nature of a human are distinct.

Yet, for the altered mouse to become a meaningful analogue for human behaviour, a second sense of nature has to be enacted, a sense encapsulated in ideas about the laws of nature. It is the assumption of similarities between species, subject to universal laws of nature, which allow analogues to be drawn between species. These are the regularities of nature, which are seen to be everywhere and always the same (Daston, 2002: 375; though see Waterton, this volume, on the precarious standardization of universal nature). Nature here is reduced to what can be shared between species and what can be held as universal, whether in clinical contexts or the behavioural genetics laboratory. The ability to transact between the two, and so use animal experimentation to reduce human suffering, is the utilitarian basis of animal experimentation in the first place. The legal status of animal bodies within laboratory spaces are reworked around this clinical translation (Asdal, 2008).

Finally, there is a third and perhaps more contested meaning of nature implied. The altered behaviour of the mouse in the video is not presented as a manifestation of its nurture; it is located in its biology, in its altered brain receptors, in its individual nature. It is innate to this animal, a manifestation of the specifically altered DNA coding for this animal's development. As Keller explains, this argument is 'specific not only to a species but to an individual, and however much it may depart from the norms of that species, it is nonetheless genetic' (Keller, 2008: 122). Nature here appears as an uneasy synonym for the genetic make-up of the individual animal. Dichotomies of nature and nurture are mapped onto a duality between genes and environment, sidelining the importance of social, historical, environmental or other experiential elements to an animal's behaviour. A form of hierarchical thinking about nature in biology, which continues to position the DNA molecule as the fundamental agent, as the natural basis of behaviour, thus emerges as remarkably resilient. It may draw on obsolete notions of genes as agential, discrete and invariant, but these endure in this experimental demonstration and in the explanatory rhetoric of the press release. As Keller suggests, this nature/nurture debate owes its perpetuity in part to slips in the use of words like 'genetic' and 'environmental' by scientists themselves, 'even while acknowledging the importance of interactions between the two, it is widely assumed that the basic division provides an appropriate starting point' (Keller, 2008: 122). Nowhere is this more evident than in media presentations of behavioural genetics as, above all, a continuing search for the genetic factors influencing affective capacities, in humans and animals, in experimental practice and everyday life.

This is, of course, not the only way of conceptualizing the nature of animal behaviour in relation to environment. Social scientists have recently returned to the history of biology, seeking more relational accounts of the sensing organism, located within its environment or *Umwelt*, drawing on the early ethologist Jakob Von Uexküll (1957), and the development of his ideas in the work of Heidegger, Deleuze and Agamben (Heidegger, 2001; Deleuze and Guattari, 1987; Agamben, 2004). Motivations for this search include the desire to replace reductionist explanations of human and animal subjectivity with accounts of interactions between organism and milieu that stress the affective and material capacities of humans, animals and environments in relation (Lorimer, 2007; Thrift, 2005). Such relational accounts promise to challenge genetic determinism, decentre the grand claims of humanism and introduce attention to the varied perceptual worlds of animals. Yet in drawing on the interpretations of ethology favoured by Heidegger, some work has been criticized for retaining a particular conception of the animal as 'poor in the world', risking reintroduction of a further anthropocentrism, where the nature of animal life is defined in opposition to the human (Elden, 2006; Acampora, 2006). Despite cautions, these arguments are worth considering, for they offer an alternative vision of the relations between human and animal behaviour, used to diagnose the implications of the changing boundaries of animality and humanity in the context of genomics.

Agamben's account (2004) starts with now well-known arguments about the behaviour of a tick in relation to the specific elements of its environment. Working in the 1920s, Uexküll proposes that the animal's behaviour can be understood through intensities of affect – the smell of butyric acid in the sweat of a mammal, the heat of a mammal's body and the textures of mammalian hair – all ultimately leading to nourishing blood vessels. The tick lives out a short life cycle that comprises, almost wholly, of an intimate relationship with these elements. The tick is open only to these environmental phenomena, 'captivated' (in Heidegger's terms) by events that trigger or disinhibit its specific behaviours. A relational ontology of the animal and its *Umwelt* emerges: the tick is this relationship; 'she lives only in it and for it' (Agamben, 2004: 47). Heidegger develops this way of conceiving of animal being in the world (Heidegger, 2001). The animal has a specific form of captivation towards elements of its environment, yet it is not conscious of this captivation. In Heidegger's formulation, the animal is caught in a relation of being openly drawn to elements in the world, yet simultaneously not exposed to the openness of being itself. As Agamben summarizes, 'the mode of being proper to the animal, which defines its relationship with the disinhibitor, is captivation. Insofar as it is essentially captivated and wholly absorbed in its own disinhibitor, the animal cannot truly act or comport itself in relation to it: it can only behave' (p. 52). The animal is captivated in its world. It behaves in the environment according to its instinctive imperatives: its nature. This nature does not reside singly in the animal or the environment, but in the relation between the two, so challenging genetic determinism, but also reintroducing a further division between the being of humans and animals.

For Heidegger, humans also share this propensity for captivation towards elements in the world. Yet human activity is distinct in its capacity to being open to being itself. We have, Heidegger suggests, moved away from captivation to being able to recognize and interrupt our relationship to captivation. As Agamben describes, 'the world has become open for man only through the interruption and nihilation of the living being's relationship with its disinhibitor' (Agamben, 2004: 70). For Heidegger, boredom is central to human experience and thus to his consideration of the open. 'Dasein [the concept Heidegger uses to question what it means to be human] is simply an animal that has learned to become bored; it has awakened from its own captivation to its own captivation' (Agamben, 2004: 70). Agamben builds on Heidegger to suggest that what is both discovered and subsequently excluded in this definition of the human is our own animality, our own captivation: 'this awakening of the living being to its own being-captivated, this anxious and resolute opening to a not-open, is the human' (Agamben, 2004: 70). The human becomes human by simultaneously recognizing and excluding this animality.

Agamben's account offers a structural connection between definitions of humanity and animality, but an asymmetry between human action and animal behaviour. However, he suggests contemporary developments in global genomics are changing this connection, seeking control over the animality of humanity. A contemporary preoccupation with understanding life and being from the perspective of the biological sciences has generated much comment from social theory (Habermas, 2003; Agamben, 2004). Much of this focuses on the challenge to human subjectivity raised by genetics and the neurosciences. For Agamben, humans, by seeking the control of life in this way and searching for the material basis of their own cognitive and emotional processes, are in the process of closing themselves to their own openness. This is couched in terms of humans making 'being' their specific disinhibitor. He suggests that the 'bare life' of animality has become the disinhibitor of human life and technology its medium, captivating humanity, whilst closing down reflections of the nature of this captivation, a process akin to animal captivation. In this context, Agamben suggests 'the total humanization of the animal coincides with a total animalization of man' (p. 77).

Agamben's work focuses on the production of ontologies of the human, and the spaces through which boundaries between human and the not-quite human are produced (see for example Agamben, 1998). It is both a political project – tracing the ontological mechanisms responsible for producing definitions of the human, and a metaphysical one – developing a philosophical gaze that reinvests meaning into life. Yet, despite the centrality of animality to his accounts of humanity, it is largely disinterested in animals themselves. Despite his identification of an 'anthropological machine' through which humanity is created with and against animality, he has little to say about the specific technological processes in which human and animal natures are bought into being (for further discussion see Oliver, 2007). Yet attention to changing patterns of captivation appears a productive analytic in the contexts of behavioural genetics, for here both patterns of animal captivation and human captivation with being are deliberately transformed. A more symmetrical account, attending not only to transformations of human nature, but also to specific animal capacities and technological interventions through which this encounter is achieved, add points from which to consider the relations between humans and animals, as well as between animals and their environment. This is the challenge as I turn to practices modelling human behaviours in the neurosciences.

Modelling human behaviour

> Dangle a mouse by its tail, and it will wriggle and strain to escape before eventually recognizing the hopelessness of its situation. Measure the time it takes to abandon thoughts of helping itself, and you have one of the classic animal tests for depression (Abbott, 2007: 6).

There is an alternative measure; it is equally provisional. In the forced-swim-test, instead of taping the tail to immobilize the animal, the mouse is immersed in water, in a straight-sided glass jar, from which it cannot escape. The researcher measures the time taken for the behaviour to change from swimming, with the goal of escaping the water, to just treading water. If the mouse stops trying to escape earlier than other mice, it is interpreted as a sign of depression. The forced-swim-test (FST) was developed in the late 1970s by Roger Porsolt. It is also known as the Porsolt test or behavioural despair test (Cryan and Holmes, 2005). Initially, it was assumed to model the acute trauma that might herald the onset of human depression. When the mouse gave up struggling in the water it was regarded as a sign of despair. The depressed mouse was seen to maintain behaviours that expressed forms of captivation towards safety and security, but this response was interrupted. Even if depression was not best understood as an interruption of captivation, this was how it was modelled in the laboratory mouse.

With the documentation of complex, overlapping and chronic symptoms in clinical depression this interpretation has changed (Craddock and Forty, 2006; LaPorte *et al.*, 2008). Scientists now acknowledge that these tests are limited analogues of human behaviour (Abbot, 2007; Cryan and Holmes, 2005). However, the forced-swim-test remains the most widely used test in researching the genetics and treatment of depression using laboratory animals.[2] From the 1970s, the test has acquired a powerful facticity in the pharmacogenetic paradigm for depression research, through an intersection between diagnostic criteria in the Diagnostic and Statistical Manual of Mental Disorders (DSM IV), the role of FDA legislation, known drug effects, established laboratory practices, and the growing use of genetically altered mice in neuropsychiatric research.

The key to the continuing validity of this model is the finding that the administration of clinically effective antidepressant treatments causes mice to engage in more active escape directed behaviours. This is particularly the case with the class of depressants known as selective serotonin reuptake inhibitors (or SSRIs). Increasing the level of serotonin available to bind to the postsynaptic receptors in mice increases the length of time these animals will struggle ineffectually to escape the water. As Abbott reflects, 'classical animal tests for psychiatric disorders are based on responses to clinically proven drugs' (Abbott, 2007: 6). This interpretative shift is underlined by Chaouloff (2007): 'the tail suspension test or the forced swimming test are by no means animal models of depression but tests which respond to antidepressants. This difference, which may appear subtle for the non specialist, is huge in terms of consequences for the construct of adequate animal models of depression'. Even if the test is now considered a limited model for human emotional disorders, its value persists as a quick and cost-effective test for the efficacy of related anti-depressant drugs.

The development of drugs in this area is challenging and expensive, 'with the cost of central nervous system drug development higher than that of any other major therapeutic area' (Cryan and Holmes, 2005: 776). However, as the rapid uptake of each SSRI indicates, potential rewards are high. With its simplicity and its predictive qualities, the forced-swim-test provides a method of rapidly screening related pharmaceuticals for new anti-depressant qualities. This approach fits well with definitions of safety and efficacy contained in US Food and Drugs Administration legislation, and the diagnostic criteria of the Diagnostic and Statistical Manual of Mental Disorders, which since the 1960s and 1970s have shaped 'psycho-pharmaceuticals into agents with specific effects' (Lakoff, 2005: 10). The forced-swim-test allows the articulation of animal, apparatus and experimenter around the known drug response. It is also a practical solution to the growing globalization of scientific research, which 'makes it safer to publish data from well-accepted tests rather than to modify them or invent new methods' (Kalueff *et al.*, 2007: 3). Through devices like the forced swim test, animal behaviour is transformed into a simplified array of standardized data, which can be circulated between laboratories.

The development of genetically altered mice, identified to express traits that perform in these tests in particular ways, has further reified both the role of simple behavioural tests and a particular understanding of genetics in modelling future treatments for human depression. The genotype of the laboratory mouse can be altered in a large number of ways; if there is interest in the animal's behavioural phenotype the battery of tests on the resulting animal is likely to include the forced-swim-test.[3] Although not a search for a genetically altered model of human depression as such, this research reinforces a view of animal behaviour understood through its genetic components. The continued articulation between forced-swim-test, drug intervention and animal model allows elaboration and refinement of the pharmaceutical interventions used to treat human depression, but this is a singular lens for studying both the meanings and mechanisms of animal and human captivation and its interruption. These 'strains of mice that show characteristic patterns of behaviour are critical for research in neurobehavioural genetics' (Crabbe *et al.*, 1999: 1670), yet they are also subject to ambiguities, which occasionally puncture these stabilized circulations of animal behaviour.

Challenging animal models

The interpretation of behavioural tests is open to question in a number of ways. Firstly, there are different versions of the meaning of the animal's behaviour measured by the forced-swim-test. Some critics argued the early quitters gave up useless struggling to save energy. As Tecott suggests, 'immobility may be alternatively viewed as a reasonably adaptive strategy for coping with this experimental situation' (2003: 652). When mice stop struggling to float on the water this could be seen as an appropriate response to an abnormal situation, rather than an abnormal response akin to human depression. Other researchers dispute this interpretation, suggesting that when mice stop struggling they display a freezing posture, a further indication of abnormal behaviour. Again there is a rebuttal: there are many interpretations of freezing postures in mice. Perhaps more subtly, there are questions about the brain processes tested through the forced-swim-test. Critics suggest divergence between drug action during this acutely stressful episode and the chronic treatment of depression in clinical settings, speculating the drug may not operate on the same brain pathways in humans and animal models (Cryan and Holmes, 2005).

There is also a range of environmental effects, which can alter or reverse genetic effects on behaviour (Würbel, 2001: 207). This was famously demonstrated to the research community through comparative research carried out when Crabbe *et al.* (1999) tested six behaviours across three laboratories. They found significant differences in the behaviours exhibited by genetically altered strains of mice, bred to express specific identifiable behavioural traits. Despite reporting going to 'extraordinary lengths to equate test apparatus, testing protocols, and all features of animal husbandry', they still conclude, 'experiments characterizing mutants may yield results that are idiosyncratic to a particular laboratory' (Crabbe *et al.*, 1999: 1670). These, they suggest, are due to often unknown environmental background effects. Despite attempts to standardize experimental procedures, there may be differences in the management of mouse colonies, in the past experiences of animals, subtle alterations in physical environments (such as light or ultrasound levels) and researcher attitudes, resulting in different outcomes in separate laboratories. In particular, there is the suggestion that varied human agencies, which exceed standardization, are intricately involved in the expression of these different animal behaviours.

The interaction between researcher expectations and animal behaviour is the focus of philosopher and psychologist Despret's discussion of Rosenthal's 1966 experiment on 'good' and 'bad' rats. In this Berkeley experiment, Professor Rosenthal presents two groups of experimental psychology students with two sets of rats, explaining that one set are inbred laboratory animals, selected to perform well in laboratory mazes, and the other are 'normal' naïve animals. Each group of rats and students performs to expectations, returning the predicted deviation in data (Rosenthal, 1966); one group perform well in the maze, the others less so. Yet, we are then told, both groups of rats are from identical albino stock; all are naïve and have no demonstrated competence with this experimental apparatus. In the relation between supervisor, students and animals the expectations of behaviour are communicated and enacted. 'The students [. . .] put their trust in their rats, emotional trust, trust that is conveyed in gestures, in students' bodies, in all these rats that were manipulated, caressed, handled, fed and encouraged: the students succeeded in attuning their rats to their beliefs' (Despret, 2004: 122).

Rosenthal had staged the experiment to seek 'the little things that affect the subjects to respond differently than they would if the experimenter had been literally an automaton' (1966: 119, cited in Despret 2004: 117). Yet, there are further questions about whether automata would be able to perform these experiments in the first place. Michael Lynch, reporting on ethnographic work in animal laboratories, suggests not. Laboratories often employ specialists who show particular abilities at 'handling animals [. . . revealing] an empathetic orientation to lab animals as living, holistic, creatures with needs to be attended and reactions to be monitored. [. . .] Scientists who are 'good with animals' can sometimes obtain compliance from their subjects which otherwise would be impossible. This is particularly the case in behavioural experimentation' (Lynch, 1988: 282). The bodily and empathetic engagement between humans and animals is vital to staging experiments in the first place. Despret questions Rosenthal's interpretation of the experiment as a search for confounding factors, his desire to eliminate these and define conditions under which the data would be uncontaminated. Instead, she argues, such attunement between bodies may be essential to the articulation of the animal body within a laboratory context; recognition of subjectivity a necessary currency for successful experimentation.

Such relational attributes make data from behavioural studies tricky to standardize and Crabbe *et al.* conclude by urging caution for the attribution of behavioural effects to genetic manipulation (1999: 1672). Yet, the main response to concerns about contingency has been to invest further in standardizing protocols, environments and practices, around shared protocols and industry standards. This ignores the additional danger that 'standardization increases the risk of obtaining results that are idiosyncratic to a particular situation' (Würbel, 2001: 210). The pressures to standardize around the simplest procedures and most economically efficient cage designs may make the behaviour exhibited more stable, but at the same time they may, ironically, make it less meaningful.

Enriching environments

Mice kept in the standard laboratory cages are known to exhibit what are called stereotypic behaviours, or stereotypies. These include cage biting, jumping, hanging, back flips and barbering, when the mice remove hair from themselves and each other. Stereotypies are repetitive, unvarying and apparently functionless behaviour patterns. In the literature, they are frequently linked to animal suffering, to be addressed though good animal husbandry (Mason and Rushen, 2006). In this literature, stereotypies are sometimes compared to human obsessive compulsive behaviours, and the likely impacts on animal welfare imputed by the distress known to be caused to psychiatric patients by similarly repetitive behaviours. However, perhaps due to a separation between animal welfare literatures and mouse genetic researchers, they are less often considered

Credit: Wellcome Images

in relation to the outcomes of behavioural tests (for an exception see Kalueff *et al.*, 2007).

To address recognized welfare issues there are now moves towards providing cage enrichments, including nesting materials, burrowing mediums and social interaction, which discourage the development of stereotypic behaviours. The types of enrichments provided are derived from ethological studies and preference tests. A recent review by Balcombe explains, these show 'mice value the opportunity to take cover, build nests, explore, gain social contact, and exercise some control over their social milieu, and that the inability to satisfy these needs is physically and psychologically detrimental, leading to impaired brain development and behavioural anomalies (eg stereotypies)' (2005: 217). Through enrichment, the stresses of captivity are to be replaced with captivation: an engaged interaction between animal sensation and the material environment, reducing repetitive and destructive behaviours.

Few mice are now singly housed, except for more aggressive males and for certain experiments. Most laboratory guidelines recommend minimum cage enrichments, including nesting materials and sometimes a cardboard tube, which can be chewed, used for shelter or as platform. The US National Institute of Health facilities recently reported 90 per cent of mice received nesting materials and 50 per cent a cardboard tube or plastic shelter (Hutchinson *et al.*, 2005). In the UK, all units surveyed provided either nesting material or sawdust substrate, with 63 per cent providing further enrichment items (Leach and Main, 2008). Such basic items are easy to incorporate within the stacked racks of standard laboratory cages. These are typically shoebox sized, approximately 30–40 cm long, 10–15 cm wide and 12–14 cm high, housing around 4 individual animals. Less widely used are larger cages, complex environments offering animals a choice of association, exercise opportunities such as running wheels, devices which encourage food searching and the variation of enrichments to support novelty seeking behaviours. Even with basic enrichments many animals continue to exhibit stereotypies. They remain extremely prevalent in certain strains – about 98 per cent of the ICR mouse strain are thought to exhibit stereotypic behaviours (Garner and Mason, 2002) – and, overall, stereotypies are estimated to afflict some 50 per cent of all laboratory-housed mice (Mason and Latham, 2004).

Much resistance to further enrichment is based on practicality, economics and past practice. However, there is opposition from some researchers. Research shows mice are able to adapt, reduce or change their behaviours using environmental resources. 'Varied environments allow animals to learn how their own actions affect their environment' (Shepherdson, 2001; cited in Balcombe, 2005: 229). Enrichment fosters behavioural competence and enhances the animal's ability to cope with the challenges of captivity; it may also alter the outcomes of behavioural tests. For Alzheimer's research, environmental enrichments significantly alter the memory data generated from genetically altered mouse models (Jankowsky *et al.*, 2005). From the perspective of the behavioural geneticist, cage enrichments introduce contingencies of environment and animal agency onto the genetic alternations used to produce behaviours and the precariously standardized tests used to measure them. This is sometimes couched in ethical terms: if enrichment increases variation in the data, more animals would be needed for each experiment, creating a conflict between refinement and reduction of animals used in scientific research. It also reveals, again, the persistence of hierarchical thinking in biology, in the search for genetic factors, before considering environmental configurations or the potential of animal agency. Thus, for now, the animals used to model human affective processes are those most likely to exhibit stereotypic behaviours. These produce the most widely circulated data, but perhaps the most ambiguous results, for the specific causes of stereotypic behaviours are still unknown.

This raises questions outside the paradigms of animal welfare and behavioural genetics, returning to philosophical reflections on human and animal captivation. As outlined above, boredom is central to Heidegger's conception of the human. There are also debates in laboratory animal studies about whether animals suffer boredom; many discussions of stereotypy use this term informally (Wemelsfelder, 1990). Anderson, writing about human boredom, suggests it can be understood in affective terms. It 'discloses a malady in the circulation of intensity' (Anderson, 2004: 744). Boredom is both in the sensing body and in the material environment (see also Anderson and Wylie, forthcoming). 'Boredom emerges once a materially heterogeneous collection of bits and pieces are held together under a type of relation that embodies the paradox that "something expected does not occur"' (Anderson, 2004: 750). Reworking boredom in terms of a suspension of affect, which is both of the body and the material world, suggests it is a relevant concept for both animals and humans. As in human boredom, stereotypies suggest a suspension of the circulation of affect involving both body and environment; they cannot be allocated simply to environmental effects or to abnormal animals. This is reflected in questioning within animal welfare literatures about 'the difference between normal animals in un-natural situations and animals which have themselves become profoundly abnormal? Between newly developed ARBs [abnormal repetitive behaviour] and ingrained habits? Between different forms of brain malfunction? As yet, we have rather more hunches than we do hard data' (Mason and Rushen, 2006).[4] Stereotypic behaviour exists in the frustrated movement between potential captivation and suspension of affect.

It is thus possible to argue for a reversal of the patterns of captivation and boredom traced by Heidegger, to follow Agamben's arguments about the changing structural connection between humanity and animality, this time through the perspective of the animal. Here again we may find 'the total humanization of the animal coincides with a total animalization of man' (2004: 77). It is humans who are captivated by the material search for understanding life – seeking the genetic basis of human and animal natures, agencies and capacities. It is the animal which exhibits the stereotypic behaviour of boredom or at least of frustrated attempts at captivation. The laboratory environment suspends the animals' immediate relationship with its disinihibitors – the bored animal becomes the contemporary model for understanding human nature. Human affective disorders are modelled through the animal, just as animal behaviour becomes remodelled through this interruption of affect.

This has implications for the future development of anti-depressant drugs and for theoretical arguments about changing relations between humanity and animality. For some researchers in psychiatry, this linking of animal models, explanatory metaphors and experimental design has limited innovation in the treatment of anti-depressants, for it is insensitive to therapeutics with different mechanisms of action. As one comment posted in *Nature* suggests: 'we have not had fundamentally new drugs in psychiatry, not because of the intrinsic inadequacy of all animal models, but because of the cookie-cutter nature of design strategies by big Pharma . . . [which] . . . has dictated one serotonergic and noradrenergic agent after another' (Watt, 2007). He goes on: 'behavioural neuroscience has been hamstrung in its efforts to avoid anthropomorphism by the complementary danger of speciesism – we think we are more different from mammals than I believe we really are' (Watt, 2007).

So does Agamben offer any way out of this scientific and philosophical quandary? Despite Agamben's interest in the exchange of properties between humans and animals, he ends *The Open* by suggesting we accept the void at the centre of our accounts of human agency and capacity. We should disconnect ourselves from the specific disinhibitor for understanding and exploiting 'being' we have found in the contemporary technosciences. Rather than seeking new forms of articulation with animality, for Agamben this means accepting a hiatus, and allowing the animal to exist outside of the sphere of being: to 'let the animal be'. As he puts it: 'to render inoperative the machine that governs our conception of man will therefore mean no longer to seek new – more effective or more authentic – articulations, but rather to show the central emptiness, the hiatus that – within man – separates man and animal, and to risk ourselves in this emptiness' (Agamben, 2004: 92). This is his challenging, but pessimistic conclusion. There is little guidance on where to go next in understanding the interplay between human and animal capacities, agencies and natures as they are enacted in behavioural genetics, or indeed elsewhere. Tracing the relations between sensing bodies, material environments and genetic technologies above provides reasons for their endurance, but also offers hints at other possibilities for relating. There is more needed than a challenge to humanism. To avoid a concomitant return to anthropocentrism 'we require a multispecies and a multi-expertise way of doing/thinking worlds and ways of life' (Haraway, 2004: 308), developing accounts and experimental practices that recognize and allocate agency in creatively open ways.

Conclusions

In seeking to recuperate from Agamben's final point of separation, I return one final time to the genetically altered mouse and the unusually compliant cat. To this image, we can now add the intermediary agencies enabling this experimental encounter and reflect on its ambiguities and exclusions anew. It is a strangely choreographed encounter of beings, unnatural even, yet it is one through which other possibilities of affective learning may emerge, which are more open than either the explanations of genetic determinism or the ethological interpretations of Heidegger suggest. The mouse may learn there are dangers from close interactions with cats; even the animals with an altered sense of smell froze when the cat meowed. The cat retains the potential to discover the consequences of playing with a costly research animal. As *The Telegraph* reports, 'even the cowardly cats would eventually begin to stare, then raise a paw, a sign they were about to pounce, at which point the researchers whisked them away from their valiant GM mice' (Highfield, 2007). This demonstrated lack of fear does not have a single author either in the animal's altered nature or in its environment – it emerges from the genetic manipulation, the prior sensual experiences of the mouse, the quiescence of the cat and its removal by researchers. The outcome of this encounter is not fixed, there is an essential openness to these interspecies associations (Morris, 2005; see also Latour, 2004; Lestel *et al.*, 2006). The project of articulating a multispecies and multi-expertise way of doing neuroscience, which is responsible to the worlds and ways of life of animals and humans is beyond the scope of this chapter, but a few starting points can be identified.

First, it requires a shift in our practices for questioning nature; a shift from demonstrating reductionist accounts to creating experiments that *add* to the world and our understanding of it (Latour, 2004). Despret suggests a difference between what is being made available and what is being rendered docile in the context of the laboratory encounter (Despret, 2004: 123). She identifies a 'contrast between the scientist who relies on the availability of both the apparatus and the animal, and a scientist who requires docility (this scientist being himself docile to the perceived prerequisites of science)' (Despret, 2004: 124). Despret calls the first caretaker, the second judge. The articulations of behavioural neuroscience traced above have tended to be reductive, requiring the docility of animal agency to known drug effects and simple behavioural tests. This is extended in the videoed encounter, the cat made docile by researchers, whisked away if the prelude of a raised paw hints at a more active engagement from this element of the mouse's 'environment'. Yet, the use of animal models can be additive and animals more active in these experimental encounters. Holistic approaches to studying mouse behaviour in their home environments, monitoring movements in space and time, recording the licks, foraging behaviour, and social interactions of their daily lives are less common, but may be more revealing. Enhancing animals' capacities can produce experiments that are more sensitive to test effects. Acting as caretaker, rather than judge, researchers breeding populations of mice in species-typical societies in naturalistic environments have enabled the detection of deficits in transgenic mice not recognized in conventional tests (Vyssotski *et al.*, 2002, cited in Balcombe, 2005: 230, see also Cryan and Holmes, 2005).

Second, it suggests a shift in our thinking about nature in the biosciences, to a view of nature as phenomenal, rather than reduction of phenomena to nature. The relational encounters which constitute this experiment suggest a multiplicity of becomings; the powers of one body enhanced or diminished only in relation to the changing capacities of another (Deleuze and Guattari, 1987). There is a blurring of the boundaries that define these bodies and the agencies that set them in motion: between natural behaviour and laboratory artefact; between the properties of DNA and drugs; between gene and environment all the way down. There is growing awareness of this complexity and multiplicity in much laboratory science. The *Nature* article for this research contains more modest queries about the way odour information activation within distinct areas of the olfactory bulb is interpreted by the brain (Kobayakawa *et al.*, 2007). The complex web of agents is expanded to encompass odour information, receptor characterization, brain function and learning. Here, we find hints of the heterogeneity of phenomenal nature, replacing the gene paradigm enduring in media presentations of genetic determinism and the impoverished understandings of animal being in the world inherited from Heidegger.

Yet these determinist accounts of species behaviour still persist in media presentation. The final addition to the videoed images is the person operating the camera, posting the images to the web, adding but also extracting from, this dance of beings. Researchers may find alternative ways of expressing the public meaning of this encounter for conceptualizing relations between genes and environment, if we still want to use those words; though perhaps here there is most resistance to alternative framings. Even as scientists recognize the limitations of the gene paradigm, Keller suggests the biotechnology industry requires this division between nature and nurture to sustain its expansionist claims (Keller, 2008). The 'diagnostic liquidity' (Lakoff, 2005) underpinning pharmaceutical development is transacted here by the exchange between known drug effects, specific measures of behaviour and genetically altered mouse models. This embodies, and links, limited models of animal behaviour and human affective disorders. Meaning and validity are sidelined; rapid circulation and persuasive demonstration is central. The video is perhaps, finally, a fitting metaphor.

This cat and mouse game is fixed. The cat will never win. Akin to the never-ending pursuit for the 'true nature' of behaviour, the mouse provides an opportunity for a captivating dance, but will remain always just out of reach.

Acknowledgements

This paper arises from a research fellowship funded by the UK Economic and Social Research Council on 'Biogeography and Transgenic Life' (grant number RES-063-27-0093). I am grateful to the ESRC for this support. I would like to thank Sarah Parry and John Dupré for the invitation to participate and other contributors for their comments on earlier drafts, particularly Celia Roberts and Richard Twine. Preliminary versions of this paper were presented to the Material Geographies workshop at the University of Durham, 19th December 2007, and to the Anthropology Department at UC Irvine, 10th March 2009. I am grateful for comments received there. I would like to thank Ben Anderson and Nicole Nelson for supplying references and draft papers, and Beth Greenhough and Jamie Lorimer for shared conversations about ethology. Particular thanks to Jamie for the reference to the fearless mouse which prompted this initial inquiry. All remaining errors are of course my own.

Notes

1 See for example, http://www.guardian.co.uk/science/video/2007/dec/12/mouse (accessed 21Apr 2009).
2 For an ethnographic analysis of experimental systems in behavioural genetics see Nelson, forthcoming.
3 This can be exemplified by a paper in the journal *Neuropsychopharmacology* (Liu *et al.*, 2007). This article explores the behaviour of three groups of mice, including wild-type mice, and the heterozygotic and homozygotic modifications to a particular gene linked to the transport of the neurotransmitter gamma-amino butyric acid in the brain. The bulk of the article involves discussion of two groups of graphs, histograms representing the response of the three mouse genotypes to the forced-swim-test and tail suspension test. In the first set of graphs, initial animal responses to these tests are represented; in the second, the measurements are rerecorded when treated with the known anti-depressant Prozac™. Wild type mice are apparently less 'depressed' when on Prozac™; yet the modified mice show little difference, opening up avenues of enquiry into specific molecular pathways.
4 On-line introduction, Mason, G. and Rushen, J. (2006) *Stereotypic Animal Behaviour – Fundamentals and Applications to Welfare* http://www.aps.uoguelph.ca/~gmason/StereotypicAnimalBehaviour/library.shtml (accessed 21 Apr 09).

References

Abbott, A., (2007), 'Model behaviour', *Nature*, 450: 6–7.
Acampora, R., (2006), *Corporal Compassion: animal ethics and philosophy of body*, Pittsburgh, PA: University of Pittsburgh Press.

Agamben, G., (1998), *Homo Sacer: Sovereign Power and Bare Life*, Stanford, CA: Stanford University Press.

Agamben, G., (2004), *The Open: man and animal*, Kevin Attell (trans.) Stanford: Stanford University Press.

Anderson, B., (2004), 'Time-stilled space-slowed: how boredom matters', *Geoforum*, 35: 739–754.

Anderson, B. and J. Wylie, (forthcoming), 'On Geography and Materiality', *Environment and Planning* A, advance online publication, http://www.envplan.com/epa/fulltext/aforth/a3940.pdf.

Asdal, K., (2008), 'Subjected to parliament: the laboratory of experimental medicine and the animal body', *Social Studies of Science*, 38: 899–917.

Balcombe, J.P., (2005), 'Laboratory environments and rodents' behavioural needs: a review', *Laboratory Animals*, 40: 217–235.

Broglio, R., (2008), 'Living Flesh: Animal Human Surfaces', *Journal of Visual Culture*, 7, 103–121.

Canguilhem, G., (1991), *The normal and the pathological*, trans. Fawcett, C.R. and R.S. Cohen, New York: Zone Books.

Chaouloff, F., (2007), comment posted in response to Abbot, A., (2007), 'Model behaviour', *Nature*, 450: 6–7, available at <http://www.nature.com/news/2007/071031/full/450006a.html> (accessed 21 Apr 09).

Collins, H.M., (1988), 'Public experiments and displays of virtuosity: the core-set revisited', *Social Studies of Science*, 18: 725–748.

Crabbe, J.C., D. Wahlsten and B.C. Dudek, (1999), 'Genetics of Mouse Behaviour: Interactions with Laboratory Environment', *Science*, 284: 1670–1672.

Craddock, N. and L. Forty, (2006), 'Genetics of affective (mood) disorders', *European Journal of Human Genetics*, 14: 660–668.

Cryan, J. and A. Holmes, (2005), 'The Ascent of the mouse: advances in modelling human depression and anxiety', *Nature Reviews: Drug Discovery*, 4: 775–790.

Daston, L., (2002), 'The moral and natural orders', Tanner Lectures at Harvard University, available at <http://www.tannerlectures.utah.edu/lectures/documents/volume24/daston_2002.pdf> (accessed 21 Apr 09).

Davies, G., (2006), 'The sacred and the profane: biotechnology, rationality and public debate', *Environment and Planning* A 38(3): 423–444.

de Chadarevian, S., (2006), 'Mice and the Reactor: The 'Genetics Experiment' in 1950s Britain', *Journal of the History of Biology*, 39: 707–735.

Deleuze, G. and F. Guattari, (1987), *A Thousand Plateaus: Capitalism and Schizophrenia*, Minneapolis, MN: University of Minnesota Press.

Despret, V., (2004), 'The body we care for: figures of anthropo-zoo-genesis', *Body and Society*, 10: 111–133.

Dumit, J., (2004), 'Drugs, Algorithms, Markets and Surplus Health', Paper presented at Lively Capital workshop, November 5, 2004.

Elden, S., (2006), 'Heidegger's animals', *Continental Philosophy Review*, 39: 273–291.

Garner, J. and G. Mason, (2002), 'Evidence for a relationship between cage stereotypies and behavioural disinhibition in laboratory rodents', *Behavioural Brain Research*, 136: 83–92.

Grimm, D., (2006), 'Mouse genetics: a mouse for every gene', *Science*, 312: 1862–1866.

Habermas, J., (2003), *The future of human nature*, Cambridge: Polity.

Hansen, A., (2006), 'Tampering with nature: "nature" and the "natural" in media coverage of genetics and biotechnology', *Media, Culture and Society*, 28: 811–834.

Haraway, D., (2004), *The Haraway Reader*, London: Routledge.

Heidegger, M., (2001), *The Fundamental Concepts of Metaphysics: World, Finitude, Solitude*, trans. McNeill, W. and N. Walker, original lectures 1929/30. Bloomington: Indiana University Press.

Highfield, R., (2007), 'Cat-and-mouse game driven by smell of fear', *The Telegraph*, 07 Nov 2007, available at <http://www.telegraph.co.uk/scienceandtechnology/science/sciencenews/3313487/Cat-and-mouse-game-driven-by-smell-of-fear.html> (accessed 21 Apr 09).

Hutchinson, E., A. Avery and S. WandeWoude, (2005), 'Environmental enrichment for laboratory rodents', *ILAR Journal*, 46: 148–161.

Ingold, T., (2000), *The Perception of the Environment: Essays on Livelihood, Dwelling and Skill*, London: Routledge.

Jankowsky, J., T. Melnikova, D. Fadale, G. Xu, H. Slunt, V. Gonzales, L. Younkin, S. Younkin, D. Borchelt and A. Savonenko, (2005), 'Environmental Enrichment Mitigates Cognitive Deficits in a Mouse Model of Alzheimer's Disease', *Journal of Neuroscience*, 25: 5217–5224.

Kalueff, A., M. Wheatona and D. Murphy, (2007), 'What's wrong with my mouse model? Advances and strategies in animal modelling of anxiety and depression', *Behavioural Brain Research*, 179: 1–18.

Keller, E., (2008), 'Nature and the Natural', *Biosocieties*, 3: 117–124.

Kobayakawa, K., R. Kobayakawa, H. Matsumoto, Y. Oka, T. Imai, M. Ikawa, M. Okabe, T. Ikeda, S. Itohara, T. Kikusui, K. Mori and H. Sakano, (2007), 'Innate versus learned odour processing in the mouse olfactory bulb', *Nature*, 450: 503–508.

Lakoff, A., (2005), *Pharmaceutical Reasoning: Knowledge and Value in Global Psychiatry*, Cambridge: Cambridge University Press.

LaPorte, J., R. Ren-Patterson, D. Murphy and A. Kaleuff, (2008), 'Refining psychiatric genetics: from "mouse psychiatry" to understanding complex human disorders', *Behavioural Pharmacology*, 19: 377–384.

Latour, B., (2004), 'How to talk about the body? The normative dimension of science studies', *Body and Society*, 10: 205–229.

Leach, M. and D. Main, (2008), 'An assessment of laboratory mouse welfare in UK animal units', *Animal Welfare*, 17: 171–187.

Lestel, D., F. Brunois and F. Gaunet, (2006), 'Etho-ethnology and ethno-ethology', *Social Science Information*, 45: 155–177.

Liu, G.-X., G.-Q. Cai, Y.-Q. Cai, Z.-J. Sheng, J. Jiang, Z. Mei, Z.-G. Wang, L. Guo and J. Fei, (2007), 'Reduced Anxiety and Depression-Like Behaviors in Mice Lacking GABA Transporter Subtype 1', *Neuropsychopharmacology*, (2007) 32: 1531–1539. doi: 10.1038/sj.npp.1301281

Lorimer, J., (2007), 'Nonhuman Charisma', *Environment and Planning D: Society and Space*, 24: 911–932.

Lorimer, J., (2008), 'Counting Corncrakes: The Affective Science of the UK Corncrake Census', *Social Studies of Science*, 38: 377–405.

Lynch, M., (1988), 'Sacrifice and the transformation of the animal body into a scientific object: laboratory culture and ritual practice in the neurosciences', *Social studies of science*, 18: 265–289.

Mason, G. and N. Latham, (2004), 'Can't stop, won't stop: is stereotypy a reliable welfare indicator?' *Animal Welfare*, 13: S57–S69.

Mason, G. and J. Rushen, (2006), *Stereotypic Animal Behaviour – Fundamentals and Applications to Welfare*, Wallingford: CABI.

McCurry, J., (2007), Japan Scientists Develop Fearless Mouse, *Guardian*, Thursday December 13, 2007 available at <http://www.guardian.co.uk/world/2007/dec/12/japan.justinmccurry> (accessed 21April 09).

Morris, D., (2005), 'Animals and humans, thinking and nature', *Phenomenology and the Cognitive Sciences*, 4: 49–72.

Nelson, N., (forthcoming) 'Blueprints for Behavior: Constructing Experimental Systems in Behavioral Genetics', PhD thesis, Cornell University.

Oliver, K., (2007), 'Stopping the Anthropological Machine: Agamben with Heidegger and Merleau-Ponty', *PheanEx*, 2: 1–23.

Rader, K., (2004), *Making Mice: Standardizing Animals for American Biomedical Research, 1900–1955*, Oxford: Princeton University Press.

Rostenthal, R., (1966), *Experimenter Effects in Behavioural Research*, New York: Appleton.

Shostak, S., (2007), 'Translating at Work: Genetically Modified Mouse Models and Molecularization in the Environmental', *Science, Technology and Human Values*, 32: 315–338.
Sunder Rajan, K., (2006), *Biocapital: the constitution of postgenomic life*, London: Duke University Press.
Tecott, L., (2003), 'The genes and brains of mice and men', *American Journal of Psychiatry*, 160: 646–656.
Thrift, N., (2005), 'From Born to Made: Technology, Biology and Space', *Transactions of the Institute of British Geographers*, 30: 463–476.
Uexküll, J., (1957), 'A Stroll through the World of Animals and Men: A Picture Book of Invisible Worlds', in Schiller, C., (ed.), and Kuenen, D., (trans.), *Instinctive Behavior: The Development of a Modern Concept*: 5–80, New York: International Universities Press.
Vyssotski, A.L., G. Dell'Omo, I.I. Poletaeva, D.L. Vyssotski, L. Minichiello, R. Klein, D.P. Wolfe, and H.-P. Lipp, (2002), 'Long-term monitoring of hippocampus-dependent behavior in naturalistic settings: mutant mice lacking neurotrophin receptor TrkB in the forebrain show spatial learning but impaired behavioral flexibility', *Hippocampus*, 12: 27–38. 10.1002/hipo.10002
Watt, D., (2007), comment posted in response to Abbot, A., (2007), 'Model behaviour', *Nature*, 450: 6–7, available at <http://www.nature.com/news/2007/071031/full/450006a.html> (accessed 21 Apr 09).
Wemelsfelder, F., (1990), 'Boredom and laboratory animal welfare', in Rollin, B., and M. Kesel, *The Experimental Animal in Biomedical Research*, 243–272. CRC Press.
Williams, R., (1983), *Keywords: a vocabulary of culture and society*, London: Fontana.
Würbel, H., (2001), 'Ideal homes? Housing effects on rodent brain and behaviour', *Trends in Neurosciences*, 24: 207–211.

Getting bigger: children's bodies, genes and environments

Karen Throsby and Celia Roberts

> We are always 100 percent nature and 100 percent nurture (Fausto-Sterling, 2004: 1510).

Any discussion of genetics is inevitably bound up in discourses of nature – a morally laden concept which is heavy with normative meaning. To do something 'naturally' (conception, childbirth, weight loss) is to do it the way it *should* be done. 'Natural', then, elides easily with 'normal'; it is, as Evelyn Keller (2008: 119) articulates, 'both a fact and a value'. But what is deemed 'natural' can also be understood as something that happens without intervention or environmental influence, facilitating the pairing of 'nature' with its alliterative opposite, 'nurture' – what Francis Galton described as 'the environment amid which the growth takes place' (Galton, cited in Keller, 2008: 122). Genetics has provided a primary site for exploring the relationship between nature and nurture; indeed, Keller describes genetics as 'the central discourse of the nature-nurture controversy' (Keller, 2008: 121), with nature/nurture forming a fundamental division that is commonly conceptualized as 'an appropriate starting point' (Keller, 2008: 122) for aetiological investigations into a range of problematized bodily conditions. Keller argues that the enduring equation of genetics with nature that underpins this 'starting point' obscures a much more fundamental question about 'what is beyond or outside of nature' (Keller, 2008: 117–24). That is, that which is deemed not natural can be allocated either to the 'non-natural' (ie 'outside or beyond nature, the physical, the internal or the spontaneously formed' (Keller, 2008: 122–2)) or to the 'unnatural' (that which 'fails to conform not simply to nature per se, but the norms (or expectations) of specific natures' (Keller, 2008: 123)). This connection to the 'unnatural', we argue, opens up a space for thinking about the normative or moral context within which demarcations between nature and nurture (genetics and environments) are drawn, and ensuing material (often clinical) interventions developed.

Keller's interrogation of what is included within 'nurture' in the nature-nurture equation signals a core feature of attempts to delineate genes and environments in disease causation – the expansiveness of the category of 'environment', which inevitably becomes *everything but* genes. This not only flattens out the different elements of environmental interaction, but also obscures the differences in amenability to change attached to those different environmental influences, the moral weight attached to different environmental factors and on whom that moral weight falls. In this chapter, following the arguments of feminist biologist and science studies theorist Anne Fausto-Sterling (2004), we ask what limitations a gene-environment distinction places on our conceptualizations of physiological 'problems'; and what the implications of refusing those distinctions (agreeing that we are always both '100% nature and 100% nurture') might be for rethinking the ways that particular problematized body conditions are approached and intervened in.

We address these questions via two case studies of problematized gene/environment interaction, described in detail below: precocious puberty and childhood obesity. As we will show, accounts of these two interrelated 'problems' rely on a gene-environment distinction: our concern here is what aetiological and moral complexities are consequently left out of these accounts? What new questions can be asked, and new kinds of interventions (or non-interventions) imagined, if the gene-environment distinction is refused?

Childhood obesity and precocious puberty:[1] what's the problem?

Puberty and obesity are two sites of human growth and development exhibiting changes at the population level that are perceived to be problematic in their own right and are considered harbingers of catastrophic futures. In the case of puberty, many argue that there has been a significant drop in the normal age of onset across the last two centuries (from 17 in 1830 to under 14 in 1960 and today to closer to 10) (Parent *et al.*, 2003: 673; Ellis, 2004: 926): a fact that is increasingly conceptualized not as an outcome of improving child health (better nutrition, more exercise), but as an indicator of wide-spread pathology, exemplified by an accompanying increase in the frequency of what is called 'precocious' puberty. Such pubertal development, occurring before 6–9 years of age,[2] is widely regarded as requiring medical treatment for several reasons; negative effects on height, increased risk of early sexual activities and drug-taking, and social difficulties associated with being physically 'out of sync' with emotional age and the development of age-peers (Parent *et al.*, 2003; Herman-Giddens, *et al.*, 2004; Palmert and Boepple, 2001). Such broad shifts in the age of onset of puberty (both 'normal' and 'pathological') are understood as indicating significant change in the hormonal conditions of human (and non-human) life (often linked to increases in exposure to toxic chemicals), raising concern about the future of human reproduction itself. Reviewing the clinical literature on puberty and menarche, for example, DiVall and Radovick (2008: 25) mention that a recent anniversary issue of *Science* magazine listed 'What triggers puberty?' as 'one of the 100 most compelling questions facing science in the next century' and note that puberty is 'a process that is necessary for the very propagation of our species'. Changes in puberty then, and particularly an increase in precocious puberty, are thought to constitute a threat to human survival.

Rising obesity rates among children (and in adults) are also seen as signalling a health crisis of epidemic proportions, with children who have reached extremes in body size held up as warnings of what lies ahead, both for them as individuals and at the population level (James *et al.*, 2001; Rossner, 2002; Wardle, 2005). The government publication *Healthy Weight, Healthy Lives: A Cross Government Strategy for England* (Cross Government Obesity Unit, 2008: x) argues that one third of children and two thirds of adults in England are currently either overweight or obese, predicting that this will rise to two thirds of children and almost nine tenths of adults by 2050. Children are identified as particular cause for concern 'because of evidence suggesting a 'conveyor-belt' effect in which excess weight in childhood continues into adulthood' (2008: 1).

Precocious puberty and obesity are linked by a number of common features. Firstly, both are understood to be the products of the interaction of genes and the environment (Weinsier *et al.*, 1998; Palmert and Boepple, 2001; Loos and Bouchard, 2003; Astrup *et al.*, 2004). In both cases, it is presumed that genetics cannot provide an adequate or total explanation for recent trends, since changes at the genetic level do not occur within such a short timescale; but neither has it proved possible, in either case, for scientists or journalists to relinquish attempts at genetic explanations. Genes cannot be blamed, yet are always invoked. As we will show, the explanatory space that this paradox creates is filled in both cases with discourses of genetic predisposition in interaction with environments. However, the specific nature of those interactions remains elusive, with genes figured as static, underlying factors that are *acted upon* by external forces in indeterminate ways.

Secondly, precocious puberty and childhood obesity not only share their roles as harbinger of future crisis, but are also conceptualized as biologically connected processes, with weight gain seen as a potential consequence of puberty, especially for girls, and excess body fat figured as a potential contributor to precocious puberty (see, for example, Ebling, 2005; Graber *et al.*, 1999; DiVall and Radovick, 2008: 24; Lee *et al.*, 2007). Furthermore, both the precociously pubertal body and the obese body are understood as bodies out of control. In both cases, this figuration is linked to particular ideas about gender- and age- 'appropriate' physical appearance and the need for normative performances of gender for children to achieve happiness and health. In particular perceived failure to achieve, or deviation from, normative standards of femininity exposes girls to significant medical interventions to 'treat' their pathologised bodies. This connection is discussed in the later part of the chapter.

Thirdly, both obesity and precocious puberty are poorly understood. Although biomedical discourse speaks confidently about *what* happens when a child goes through puberty, little is known about what initiates those changes, in both 'normal' and 'pathological' cases (Ebling, 2005; Hermann-Giddens, Kaplowitz and Wasserman, 2004; Gianetti and Seminara, 2008; DiVall and Radovick, 2008). Similarly, the bodily mechanisms of weight loss and gain are poorly understood; an effective understanding of mechanisms of appetite and satiety, for example, remains elusive (Shell, 2002; Loos and Bouchard, 2003; Astrup *et al.*, 2004; Gard and Wright, 2005). These shared uncertainties provide a further basis for interrogating the complexity of the relationship between the conceptual domains (biology, society, nature, genetics) within which 'getting bigger' is negotiated scientifically.

However, there are also important differences between obesity and precocious puberty that create openings for rethinking the relationships between bodies, genes and environments in, and across, both contexts. Firstly, while both puberty and obesity are seen as processes of becoming, and while the environments that are seen as creating both of these 'problems' are social, in puberty discourses, biological and metabolic *processes* remain central. Obesity, on the other hand, is conceptualized not as a bodily process, but as a problematic bodily state (that can be 'diagnosed' via a simple calculation of kg/m^2). Puberty, then, is something that we *go through*, while obesity is something that we *are*. Popular discourse on obesity, as a result, tends to jump from behaviours (consumption/inactivity) to obesity (the problematized bodily state), without focusing on the processes (such as metabolic responses) which the body undergoes whilst getting bigger. However, it is also the case that within the medical context, obesity is increasingly conceptualized in metabolic terms – for example, in the recent re-naming of the British Obesity Surgery Society as the British Obesity and Metabolic Surgery Society[3] – and research focused on the bodily process of weight gain, with the goal of producing mechanism-specific drug interventions (see, for example, Bray and Tartaglia, 2000; Lichtenbelt *et al.*, 2009).

Secondly, precocious puberty and childhood obesity are framed as very different kinds of 'problems'. In puberty, 'normal' development is understood as largely genetically determined (some are even prepared to put a figure on this: Wehkalampi *et al.* (2008) argue that 86 per cent of variance of pubertal timing in girls is genetic, compared to 82 per cent for boys), whilst 'abnormal' puberty is seen as caused by the action of the environment – which could mean anything from nutrition to endocrine disrupting chemicals to the absence of a father figure – on those 'normal' genes and gene pathways (Parent *et al.*, 2003; Banerjee and Clayton, 2007; Gianetti and Seminara, 2008; Roseweir and Millar, 2009). In this sense, in precocious puberty, 'normal' gene pathways or cascades (the most frequently cited being the kisspeptin/ KISS1R system) are disrupted by external influences. In contrast, obesity is commonly conceptualized as resulting from an undesirable genetic propensity to fatness that is enabled by environmental and social factors. Weinsier *et al.* (1998: 148) describe this in terms of 'previously silent' genetic variants becoming manifest: 'Simply put, our genes permit us to become obese; the environment determines if we become obese'. This is linked to the third key difference between precocious puberty and obesity – the moral freight that each condition brings with it, both for the individuals involved, and in the case of children, for those responsible for their care.

When precocious puberty occurs, it is conceptualized as a tragedy for individuals, but one for which those caring for them are not responsible (see, for example, O'Sullivan *et al.*, 2002). If the environment is to blame, this is not an environment that is understood as controllable by adults in any 'simple' way.[4] Conversely, obesity in children is seen as the result of the moral failure of adults, although blame can also be seen to be falling increasingly on children, who are exhorted to take individual responsibility for their size (see, for example, Evans *et al.*, 2003; Evans *et al.*, 2004). The opposing discourses of victimhood (the tragic disruption of 'normal' gene pathways in precocious puberty) and moral failure (the enablement of genetic propensities towards obesity) underpin the differential moral framings of the two 'problems'. This distinction also highlights the fact that it is the *timing* of puberty that is seen as problematic, rather than puberty *per se*. Obesity, on the other hand, is *always* conceptualized as problematic and undesirable, with the focus on children positioned as an attempt to head adult obesity off at the pass, rather than to see children through a particular stage in their development. These differences in framing signal the different freight of individualized responsibility that each condition carries and raise questions about how such freight might be better acknowledged and accounted for in medical and scientific research, as well as scientific and popular discourses. In the final part of this chapter, we suggest a theorization of gene/environment systems that allows such questions to be centrally positioned rather than sidelined.

Following the genes and gene pathways in our two case studies

Although, as argued in the introduction to this volume, simple 'gene for' claims are rarely considered sustainable in contemporary genetics, in our two case studies the search for a genetic basis has not been abandoned, with considerable effort and funding being invested in establishing the mechanisms and proportions of genetic influence and the relationships between different genes and/or gene pathways and particular environments. Twin and adoption studies are frequently cited as evidence of heritable genetic predisposition to obesity; Friedman (2004: 563), for example, argues that 'the heritability of obesity is equivalent to that of height and greater than that of almost every other condition'. However, the reported heritability of obesity from these studies varies enormously. Loos and Bouchard (2003: 403), for example, found estimates ranging from 5 per cent to 90 per cent, with study design and the kinds of families studied generating dramatically varying results. Nevertheless, they are unwilling to relinquish genetically determined obesity, listing multiple 'candidate genes', and setting out four categories of genetic determination for obesity that reflect different degrees of gene-environment interaction: genetic obesity (rare monogenetic disorders), strong genetic predisposition, slight genetic predisposition and genetic resistance to obesity/obesogenic environments. A very small number of monogenetic obesity disorders have been identified, but these are very rare (<0.1% of the population according to Frayling *et al.* (2007: 889)) and are therefore unable, according to Loos and Bouchard (2003: 409) to 'explain the magnitude of the obesity problem that industrialized societies are facing today'. Instead, as is evident in the term they use – 'polygenic/common forms of obesity' (p. 409) – most obesity is seen to be aetiologically complex. This recognition of complexity drives a search to identify multiple candidate genes and their interactions with other genes and environments. According to the 2005 update to the Obesity Gene Map (Rankinen *et al.*, 2006), 253 human obesity quantitative trait loci (QTL)[5] have been identified (up from 208 in 2003 (Snyder *et al.*, 2004), and just 44 in 1999 (Chagnon *et al.*, 2000)), marking the increasing scientific focus on identifying the genetic pathways through which obesity develops. In 2007, Frayling *et al.* published a paper in *Science* announcing the identification of the FTO (fat mass and obesity) gene, stating that the 16 per cent of adults who are homozygous for the risk allele in their study weighed approximately 3 kg more than those without it. This was taken up excitably in the press, signalling a strong public appetite for these genetic explanations: 'Obesity is not just gluttony – it may be in your genes' announced the *Guardian* (Randerson, 2007); and, less moderately, in the *Daily Mail*: 'Proof: There IS a gene that makes you fat' (McRae, 2007). The research team, themselves, however, reflected far more uncertainty in their own assessment of their discovery, pointing out that 'FTO is a gene of unknown function in an unknown pathway [. . .]' (Frayling *et al.*, 2007: 893).

In early puberty, one particular gene pathway has received much attention since being identified in 2003 as critical to the initiation of normal puberty. This pathway, involving the hormone kisspeptin and its receptor gene KISS1R, has been extensively studied in mice, monkeys, sheep and in particular small groups of humans (case studies of individuals or families experiencing 'pathological' puberty), with some arguing that KISS1R can be understood as 'crucial' in relation to puberty (Gianetti and Seminara, 2008: 299; Roseweir and Millar, 2009), whilst others place it only amidst a longer list of candidate genes (DiVall and Radovick, 2008: 24). Although the kisspeptin/ KISS1R system might be necessary for normal puberty to occur, this gene pathway cannot explain either normal or abnormal puberty: as Messager (2005) argues, it remains unclear what causes the initial secretion of kisspeptin in the brain. Despite this, media accounts of this work stated that 'A team of researchers . . . *recently pinpointed a gene* that they believe controls puberty through the regulation of a protein called GPR54 [the earlier name of kisspeptin]' (Roberts, 2005, emphasis added) or, more lightheartedly, described kisspeptin as the 'chemical kiss that turns kids into adolescents' (McKie, 2005). 'It is sealed with a kiss,' one report stated, 'Researchers have found that a protein called kisspeptin triggers the cascade of biochemical changes that leads to puberty and turns children into hormonally challenged adolescents' (McKie, 2005).

In both conditions, 'genes' or 'genetic predispositions' are thought to operate in interactions with each other and/or the environment (including metabolisms). In obesity research, this fact is repeatedly cited as an *obstacle* to effective testing for a propensity towards the condition. Marti *et al.* (2004: S29) argue, for example, that 'The occurrence of gene x gene and gene x environmental factors interactions makes it more difficult to interpret the specific role of genetics and lifestyle factors'. Complexity and interaction here are figured as muddying the waters, obscuring underlying truths and certainties which are deemed to be knowable in the future. Indeed, one of the common features of the articles we reviewed is a caveat about the limitations of what can be known from a particular research finding and the need for further research (see, for example, Frayling *et al.*, 2007; Marti *et al.*, 2004).

These models of gene-environment render individuals genetically susceptible to pathology *in some environments*. Sometimes, this is interpreted as having the '"right" genes in the "wrong environment"' (Rosmond, 2004: 180), as in the case of the so-called 'thrifty gene' – a presumed genetic propensity for the body to store fat as insurance against famine (Neel, 1999; Spiegelman and Flier, 2001).[6] Indeed, Hill and Peters (1998: 1371) suggest that increases in body mass in order to restore energy balance could be seen 'not as a result of defective physiology, but as the *natural* response to the environment' (emphasis added). This model positions genes and environments as fundamentally mismatched, mirroring the nature-nurture dichotomy discussed above, with the 'natural' response figured here as appropriate, and even positive. From a different perspective, a 'faulty' gene is given expression through exposure to environmental factors; as Kopelman (2000: 635) suggests, 'implicit to the susceptible gene hypothesis is the role of environmental factors that *unmask* latent tendencies to develop obesity'.[7]

In each case, 'genes load the gun, but environmental factors pull the trigger' (Perrin and Lee, 2007: 308), either by enabling an undesired propensity or by being unable to constrain it. The action (or the potential action/expression) of genes is figured as static – a 'fact of nature' to be discovered ('unmasked') – while the environment ('nurture') is, almost by definition, changeable. This presumed changeability of the environment renders it the obvious site of intervention in response to the 'problems' of both precocious puberty and childhood obesity (and obesity in general). In obesity, in the absence of any viable genetic intervention (apart from genetic tests to detect the presence or absence of a particular gene or gene sequence) the conclusions of papers habitually fall back on the significance of so-called 'lifestyle' changes in diet and exercise oriented both towards the achievement of weight loss, or especially in the case of children, the prevention of weight gain (see Gard and Wright, 2005: ch. 7). John Hewitt, for example, in his 1997 paper on obesity asking, 'What have genetics studies told us about the environment?' concludes that 'Genetic studies are helping us to refine our understanding of environmental risks and to focus preventive or ameliorative efforts on those categories of risk that are most likely to be important' (1997: 357). In another example, Friedman (2000: 634), while arguing that 'tremendous scientific opportunities abound', asserts that 'for the moment, there is no panacea. So the final message is this . . . weight loss and exercise improve health', although with the recognition that 'a robust biological system makes it exceedingly difficult for most individuals to maintain weight loss for an extended period of time' (Friedman, 2000: 634). Indeed, Speakman (2004: 2094S) suggests that 'Addressing the genetic side of this interaction is potentially a far more tractable problem than addressing the environmental component by re-engineering society, because the level at which interventions might ultimately be made is the individual rather than society as a whole'. However, in the absence of a feasible, permanent intervention at the genetic level – 'the ultimate 'holy grail' solution' (p. 2102S) – he is forced to return to the more prosaic strategy of drugs (which are not yet available, but which we are 'likely to see') 'at the same time as calorie restriction' (p. 1021S).[8]

Similarly, in precocious puberty, the site of intervention has to be 'the environment', or at least the hormonal elements of the physiological processes involving genetic cascades; treatments for precocious puberty involve drugs which intervene in neuroendocrinological processes and prevent the cascading effects of hormones and gene pathways on developing bodies. Decisions to prescribe these medications are complex, because unlike obesity (which is never considered normal or acceptable and is therefore always open to intervention), 'normal' and 'precocious' puberty are slippery terms (puberty is ultimately desirable, but needs to happen at the 'right' time) and the decision to treat is made on a case-by-case basis. As for childhood obesity, however, genetic knowledge provides no direct avenue for intervention: the promise, as discussed in more detail below, is only to detect the children most at risk.

What kinds of intervention are possible?

Technoscientific claims about the proportion of influence accorded to genes versus other factors in obesity and early puberty are not simply a matter of intellectual argument, then, but constitute a debate that has serious implications for action. Although notions of genetic causes rarely lead to interventions at the level of the gene, the kinds of other interventions developed and offered are dictated by scientific beliefs about cause. If genes are implicated in both childhood obesity and pathological puberty, the focus of hope is to identify particular genes or gene sequences in order to find those individuals who are more likely to suffer from the condition, even before any evidence of 'symptoms'. The hopes attendant on such early identification is that intervention could prevent the onset of conditions that are otherwise very difficult to treat, allow for scarce resources to be allocated most effectively, and produce individualized preventative strategies.

For children understood to be at genetic risk of precocious puberty, this dream is strong: in part because it is much easier to prevent puberty with hormonal drugs than it is to slow it down or reverse it once it has started. The clinical literature points to the benefits of this procedure for both children and parents, who would ideally be saved the trauma of unusually early development. Phillip and Lazar, for example, in their paper 'Precocious Puberty: Growth and genetics' (2005: 56), argue that 'detection of these genes will provide a tool for identification of children at risk of developing CPP [central precocious puberty],[9] enabling early intervention with the aim of preventing its distressing outcomes.' All of this work leaves aside the question of the psychological effects produced by the interventions themselves, figuring medical treatment as preventing distress, in contrast to the traumatic experience of precocious puberty. Whilst this is an empirical question yet to be explored (that is, the question of whether taking hormonal medication to prevent the onset of puberty creates psychological problems for children or simply alleviates them), there is evidence in related fields that such intervention may be harmful in both the short and long term. A recent study of adult women who were treated as children with hormones to prevent unusual ('pathological') height has shown that many of these women experienced this medication intervention (and its attendant examinations) as traumatic, and indeed, were more likely as adults to suffer from depression and anxiety (Rayner, 2009). The literature on surgical and hormonal interventions in babies and young children diagnosed with forms of intersexuality also shows that such interventions are rarely experienced as neutral, and can be viewed by children and by adults looking back to their childhoods as traumatic and psychologically destabilizing (Creighton and Minto, 2001).[10] The dream of early intervention, then, may 'make sense' biologically, but needs to be thought through in terms of its psychological consequences.

As argued above, childhood obesity is conceptualized as a problematic state rather than a process, and increasingly, as a state that is highly resistant to change. In this case, the potential of targeted intervention subsequent to genetic testing is more open-ended: children who are already overweight could be tested and targeted for clinical interventions, and those who are deemed genetically 'at risk' could become the focus of preventative strategies. As Perusse and Bouchard argue, 'definition of these interaction effects for phenotypes related to obesity is therefore important because it will eventually allow the identification of individuals at risk of obesity, the development of complications associated with obesity, and the identification of those likely to be resistant to dietary interventions and hence requiring, perhaps, more drastic or better-adjusted dietary prescription' (2000: 1285S). At the level of the individual, the psychosocial effects of this surveillance and intervention are not addressed within these recommendations. This is in spite of mounting evidence that schemes such as the routine weighing and measuring of schoolchildren (Colls, 2007; Evans 2007), and healthy eating / weight management educational programmes (Evans *et al.*, 2003; Evans *et al.*, 2004) disturb children's (and particularly girls') self-esteem and confidence in relation to their bodies and food, as well as producing physiological disruptions as children enter into the cycles of weight loss and regain that characterize the histories of many who struggle to maintain a 'healthy weight'. It is also difficult to distinguish these risk-based interventions from those already in place for weight loss; indeed, one of the research team working on FTO commented in the *Guardian*, 'When you have this risk factor or not, if you are overweight, you should eat less, exercise more' (Randerson, 2007). However, this 'rational prescription' (Ogilvie and Hamlet, 2005: 1545) ignores the fact that over 80 per cent of all weight loss interventions are unsuccessful in the long term (Mann, Tomiyama *et al.*, 2007). The recommendation to eat less and exercise more presumes the efficacy of this as a sustainable weight loss / management strategy, even in the face of overwhelming evidence that this is not the case. Indeed, radical dietician Lucy Aphramor (2005) argues that this is an unethical recommendation for doctors to make, since no reliable means currently exists to enable people to achieve it, exposing them to the risk of further failure and blame for their pathologized condition.

The persistence of normative elements: moral concerns around getting bigger

Discussions about the differing roles of genes and environments in childhood obesity and precocious puberty, at least currently, have little effect on the treatments offered to children. We are tempted to ask, with Keller (2008: 117), 'why, given all the conceptual difficulties in attempting to sort nature from nurture, do we persist?' Why do scientists expend so much effort attempting to make distinctions between genes and environments and quantifying their respective contributions to the production of embodied differences?

Keller's response in 'Nature and the Natural' (2008) is instructive. After describing the ways in which contemporary technoscientific discourses commonly take it 'for granted that the organism can be regarded as a sum of genetic and environmental factors, with interactive effects constituting a persistent demarcation' (Keller, 2008: 122), she suggests that the ubiquitous discursive aligning of 'genetic' with 'natural' (and therefore both 'usual' and 'legitimate') means that genetic investigation has a role in every attempt to produce knowledge about human (and other living entities') behaviour. Genes, it is assumed, can explain both the normal and the pathological: even whilst, as we have argued in our two case studies above, 'environments' are thought to matter (somehow), it is only insofar as they 'interact' with genes that they are considered powerful. Similarly, it is the presumption of the power of genetics to explain both normality and pathology that renders the possibility of successful intervention at the genetic level, to use Speakman's term, the 'ultimate "holy grail" solution' (2004: 2102S). It is this paradox – that genes will explain everything but yet are only ever part of the picture – Keller argues, that makes genetic science so powerful:

> No wonder then that, today, it has become so easy for genetics to subsume so much under the rubric of nature – not only the functional and the dysfunctional, the normal and the pathological, the expected and the unexpected, but also the entire gamut of

> human behaviour, from sexual preference to political choices under the same rubric. All this while at the same time maintaining an opposition between nature, nurture and the environment, nature and culture, perhaps needing that opposition in order to stake their expansionist claims (Keller, 2008: 124).

Thinking of our two case studies in these terms and within the current climate of intense concerns about obesity epidemics and the decrease in pubertal age across the western world, we suggest that the unremitting focus on genes is indeed part of an expansionist claim to know and potentially control unruly bodies. Claims to know obese and unnecessarily pubescent bodies are powerful in this climate because these bodies are matters of serious moral concern in medical, scientific and public discourses. These concerns take different, but related forms in our two case studies, with moral failure seen as causing and sustaining obesity, whilst being a likely consequence of precocious puberty. As Throsby (2009) argues elsewhere, public discourses on obesity constantly produce fat adult bodies as a threat to society: a moral failing of the will to health that is part of today's citizenship duties. Obese, or potentially obese children's bodies, are consequently harbingers of doom and sites of even more intense moral approbation: of course the child cannot be held responsible for his or her failure to maintain a 'normal' weight, but those in charge of his or her eating certainly can. Despite claims about genes, then, the parents of obese children are figured as morally reprehensible failures: unable to assimilate widely-available information about nutrition, unable to control their children's intake of food, or unable to understand the long-term consequences for the child and society more broadly of being overweight. The moral panic (Cohen, 2002) around obesity, particularly as it relates to children, has thrown up a predictable cast of contemporary folk devils – perhaps most famously in the United Kingdom, the 'Burger Mum', Julie Critchlow, who attempted to undermine celebrity chef Jamie Oliver's attempts to reform the school dinners being served at her children's school by passing burgers and chips through the fence and was subsequently labelled by Oliver himself a 'big old scrubber' (Martin, 2008). However, the singling out of individuals or groups (working mothers, McDonalds etc) is underpinned by a much more established and pernicious moral discourse around fatness – around bodies which are deemed to be out of control, disregarding of their own health, lacking in personal hygiene, lazy, unintelligent etc. (Cooper, 1998; Wann, 1998; Gard and Wright, 2005; Murray, 2005). These presumed characteristics of fatness are readily reproduced by children, who learn from an early age the unacceptability of the fat body, and by implication, the need to exercise discipline over it in order to exert change (Latner and Stunkard, 2003). In this light, the very presumption of the changeability of the environment/lifestyle (as opposed to the intractability of genes) reproduces these pejorative discourses of blame.

Significantly, several of the papers reviewed in preparation for this chapter address directly the issue of the stigmatization of obesity, arguing that genetic knowledge absolves individuals of blame for their condition. Friedman, for example, concludes that 'We can only hope that advances in our understanding of the causes of this condition will lead to changes in the perception of what it means to be obese in a world of harsh judgments and facile conclusions that are not supported by a growing set of scientific facts. The stigma of obesity should be discarded, enabling this disease to join with other conditions that required that we look beneath the surface' (2004: 568).[11] But however good the intentions of these exhortations to abandon the stigmatization of obesity, these claims maintain the status of obesity as a problematic condition about which something can, and imperatively, should be done. It may not be an individual's fault that they are obese (because of their genes), but the obligation to act remains, even in the absence of a safe and effective intervention. This renders the failure to act morally problematic, regardless of genes or genetic predispositions. Furthermore, issues of body size, and the pressure to act in relation to those issues, are always profoundly gendered, with women confined by much narrower parameters for what constitutes acceptable body size and subject to far greater surveillance in relation to their weight than men; indeed, they already make up the vast majority of consumers of every weight loss intervention (Cooper, 1998; Stinson, 2001; Wann, 1998).

In precocious puberty, moral concerns focus on the presumed *sequelae* of early development: precocious sexual behaviour and higher risk of substance abuse. In the clinical and popular literature (in newspaper reports, for example), discussions of early puberty are often framed by worries about (female) sexual precocity and its potential consequences. A recent article in the *Guardian's* 'Family' weekend supplement, for example, summarized research that reportedly found that 'entering puberty young (before 11) correlates with a host of problems, from teenage pregnancy to depression. Only 2 per cent of those who do so go on to enter higher education, regardless of their parents' IQ and educational level' (James, 2009: 2). Another headline makes the point more dramatically: 'Girls With Early Puberty, Older Boyfriends At Greater Risk For Drugs, Sex, Alcohol' (Anon, 2007), with the article arguing that girls who mature early are more likely (than boys who mature early) to have older romantic and sexual partners and to take illicit drugs. The risk posed to girls of entering puberty very young is, many claim, that they will be taken for sexual subjects before their time, and, even worse, may consider themselves sexual subjects and behave 'inappropriately' for their age: 'Early maturing girls are more likely to be unprepared emotionally and cognitively for the changes of puberty itself *and the social pressures that they may experience from boys and adults who perceive them to be more mature than their age and experience*' (Graber *et al.*, 1999: 110, emphasis added).

Although in both the clinical and popular literatures, the dangers of such precocity tend to be assumed rather than spelled out or evidenced (see for example, Moshe and Lazar, 2005: 59; Posner, 2006: 319), these concerns are sometimes linked directly to recommendations for treatment. In an article on the use of drugs affecting the secretion of a particular hormone in the brain to treat precocious puberty, for example, Mul and Hughes (2008: 5) describe the problems caused by precocious puberty constituting indications for such treatment. After describing height 'impairment' (children who go through puberty early tend to end up shorter than their peers), and before advocating the use of hormones to 'halt menses for hygienic reasons in some girls with severe cerebral palsy or developmental retardation' (such girls tend to go through puberty early), they mention the 'psychosocial outcome[s]' associated with precocious puberty:

> Few studies have evaluated psychosocial outcomes following early or precocious puberty, but a long-term Swedish study reported more antisocial behaviour in adolescence and lower academic achievement in adulthood. . . . *There is evidence that normal, early maturing adolescents are more likely to have sexual intercourse and engage in substance abuse at an earlier age than normal or late maturing adolescents* (Mul and Hughes, 2008: 5, emphasis added).

This statement is left hanging, and the reader can only assume that these 'psychosocial outcomes' of puberty are, like 'hygienic problems' and 'height impairment', considered suitable indicators for treatment.

Although their paper tends to indicate that the use of hormones to treat precocious puberty might prevent early sexual activities, Mul and Hughes (2008: 6) ultimately acknowledge that 'embarking on treatment for the individual child with CPP [central precocious puberty] is not underpinned by any comprehensive evaluation of the long-term outcome data at a population level.' Despite this lack of evidence, the emphasis on the potential of such treatments to produce psychosocial outcomes are highlighted in their conclusion. Such effects are seen to be 'equally important' to the physical effects pertaining to height and menstruation:

> The aims of the treatment are to halt and perhaps reverse the secondary sex characteristics of puberty, prevent early onset of menses in girls and attenuate the loss of height potential consequent upon advanced skeletal maturation. These are the physical goals of the treatment intervention, *but the psychosocial aspects are equally important to enable the child and family to cope better with the 'mismatch' between physical and emotional development* (Mul and Hughes, 2008: 7, emphasis added).

Psychologist Rachel Posner (2006) makes a similar claim about research into early puberty, arguing that it is imbued with assumptions associated with 'societal anxiety about adolescent female sexuality [which] has inhibited young women's ability to healthfully and positively consolidate their sexuality into their identities' (Posner, 2006: 316). Indeed, she asserts in her conclusion that 'underlying the agenda of most of the research into the antecedents and consequences of early puberty has been a drive to control and limit adolescent girls' sexuality. This line of research,' she continues, 'has aimed to determine how sexual behaviour among adolescents can be prevented' (Posner, 2006: 320). Concerns about girls entering puberty early also focus on cultural judgements about the aesthetic qualities of bodies. Posner writes that,

> Perhaps one of the most difficult challenges facing early maturers is that they do not fit the cultural image of ideal feminine beauty, that of a tall, slim figure. Girls who

> matured earlier were bigger than their peers and often grew to be heavier and shorter than their peers (Posner, 2006: 319).

There are obvious links here to concerns about obese girls, for whom failing to meet contemporary aesthetic standards is also seen as having serious psychological consequences. Indeed, Posner suggests that it is difficult to separate the effects of these conditions on girls' self esteem, as they are so often experienced together: 'Further research is needed,' she writes, 'to disentangle the psychological effects of obesity from the effects of early puberty. That is, are girls concerned about their figures because they are early maturers or because they are overweight?' (Posner, 2006: 319).[12]

In both our case studies, gender is at stake in the moral framings of the conditions and their presumed consequences. Moral concerns focus on girls in particular, and, involve gendered assumptions about aesthetic qualities of bodies (thin is better), sexuality (girls should not express a sexual interest too early) and responsibility. In relation to the latter, in puberty discourses, if young girls excite inappropriate sexual interest, they – rather than those who display such interest – must be changed. Obese girls, similarly, must learn a gender-appropriate concern about their eating and exercise habits and learn to take responsibility for their size. In every case, we want to suggest that there could be alternative framings of the 'problems' caused by the conditions under discussion: framings that as we explain below, focus not on the individual, pathologized body, but on the social world constituting such pathologies

Conclusion: From interaction to relation?

In a careful analysis of the scientific literature on the genetics of 'race', Anne Fausto-Sterling (2004) argues that the contemporary focus on genetics as a means to explain the stratified epidemiology of particular health problems (heart disease, high blood pressure and diabetes, for example), assumes that diminishing such health inequalities cannot be done via other routes. What is papered over in the attempt to discover genetic bases for the racialized distribution of these conditions, she contends, is that basic health measures would have far greater impact on reducing morbidity and mortality (Fausto-Sterling, 2004: 24). Thus, 'it is not what the new biology of race will produce,' that constitutes the key risk of this field, 'so much as it is that the new biology of race diverts our attention from solving problems using solutions we already have at hand' (Fausto-Sterling, 2004: 30).

For Fausto-Sterling, this argument is linked to her proposal, made in a later paper on bone health and sex/gender, to retheorize the relationship between 'genes' and 'environments', understanding the actions of genes within complex systems. Rather than as causes, she suggests, 'genes are best understood as mediators suspended in a network of signals (including their own) that induce them to synthesise new molecules' (Fausto-Sterling, 2005: 1507). Genes, in this view, 'do not build organisms from the bottom up; rather, their activities are sandwiched somewhere in the middle of chains and networks of events that integrate organisms with their environment' (Fausto-Sterling, 2004: 26). In the case of racialized differences in health, she argues that 'it is not that different biological processes underlie disease formation in different races, but that different life experience activates physiological processes common to all, but less provoked in some' (Fausto-Sterling, 2004: 26). Our bodies, in other words, 'naturally' produce responses to our environments (Fausto-Sterling, 2004: 31): living as a black person in a racist US culture, for example, leads to increased blood pressure for many individuals.

In contrast to older scientific notions of homeostasis (where bodies are understood to be constantly trying to achieve a 'normal' state), this model, developed by physiologists in the late 1980s, argues that bodies are allostatic, functioning as a kind of '"smart" thermostat' that regularly changes activity to meet anticipated demands (Fausto-Sterling, 2004: 27). This view of the body takes into account the social experiences of organisms (including humans) in understanding the development of particular disease conditions. In relation to high blood pressure, for example,

> The allostasis model does not ratchet up the hunt for a 'broken' gene to explain essential hypertension. Instead, it proposes that hypertension is an orchestrated response to a predicted need to remain vigilant to a variety of insults and danger – be they racial hostility, enraging acts of discrimination, or living in the shadow of violence. Over time, all of the components that regulate blood pressure adapt to life under stress (Fausto-Sterling, 2004: 27–8).

For us, what is compelling about this proposal is the challenge it poses to more prevalent models that attempt to separate and quantify the different contributions of 'genes' and 'environments', and, in particular, what this challenge means for health-related interventions. In the case of hypertension, Fausto-Sterling argues that the allostasis model emphasizes what she calls 'higher-level' interventions, adjusting entire systems of physiology and attending to the emotions and unmet needs that produce the negative physiological results of 'hypervigilence' around blood pressure (individuals' brains, she argues, can become extremely sensitive to changes in blood-related systems). The overall aim of interventions, she argues, should be to 'reset the body's response systems, allowing function in a range less likely to produce excess morbidity and mortality' (Fausto-Sterling, 2004: 29). In this case, citing neurophysiologist Sterling (2004), Fausto-Sterling approvingly lists 'lifestyle interventions' (dietary changes, weight loss, exercise and reduced alcohol consumption) as potentially leading to such systemic change, in part (in contrast to drug therapies) through acknowledging need and 'enlarging positive social interactions and revivifying the sense of connectedness' (Fausto-Sterling, 2004: 29–30).

While we are encouraged by her willingness to look to different kinds of interventions from the conventional drug therapies, we are less convinced that such interventions will have much effect in our examples. Indeed, part of what we have been critiquing here, particularly in the case of childhood obesity, is the fact that even in the wake of immense interest in genes and failure to demonstrate that 'lifestyle' interventions actually work in the long term, recommendations to 'eat less and exercise more' still retain most-favoured status in terms of intervention. Indeed, we have argued that such emphasis can have precisely the opposite effect to 'enlarging positive social interactions and revivifying the sense of connectedness': they can reinforce and increase experiences of discrimination and alienation. Although the case of early puberty (in which drug treatments are the normal recommendation) might hold closer parallels to hypertension (hence mirroring Fausto-Sterling's recommendation to move away from drugs to more 'lifestyle' oriented interventions), we would likewise suggest that what might be most important for children diagnosed with this condition is to simply avoid conceptualizing their puberty as a disaster.

To make effective interventions into the so-called epidemic of childhood obesity and the purportedly population-threatening trends towards early puberty, we need to think seriously about issues pertaining to economics, gender, class and ethnicity at both macro and embodied levels. Taking an allostatic model seriously means thinking about bodies in relation and consistently refusing a separation between genes and environments (even the separation assumed in most interactionist models). It also means challenging cultural norms around the moral freight and personal responsibility associated with particular bodily conditions. As our two case studies show, investigations into the genetic causes of conditions or attempts to delineate gene/environment interactions in producing them tend neither to reduce either the moral freight of conditions nor to shift already-existing attributions of responsibility. Parents of children diagnosed with obesity are still more likely to be held responsible for this than those whose children are early developers. Thinking of ourselves as 'always 100 percent nature and 100 percent nurture', as Fausto-Sterling suggests, means that moral questions cannot be separated off from physical ones, or left to be answered later. Cultural norms and practices associated with gender, class, ethnicity and other forms of body-based discrimination become sites of intervention that may well have lasting impacts at both individual and population levels.

Notes

1 For reasons that will become clear throughout the chapter, we dislike both these clinical terms, which carry a weight of complex social opprobrium that we are not keen to reproduce here. Rather than litter the text with inverted commas, however, we use them with the hope that the reader will sense a kind of distancing throughout.

2 This range of numbers reflects differences in age ranges for pathology proposed in one key article, according to gender and ethnicity. For white girls, the authors suggest, puberty should be considered pathological if it occurs before 7–8, for white boys, age 9. For black girls, the proposed age range is lower (6–7 years), with no additional guidelines being specified for black boys (Parent *et al.*, 2003: 675). Mul and Hughes (2008: 3) draw a distinction between 'early' puberty (that occurring at ages 8–10 for girls and 9–11 for boys) and 'precocious' puberty (that occurring earlier than these ages). In terms of treatment, however, this is not a functional distinction: children with early puberty may also be treated with hormonal medications (described in more detail below).

3 http://www.british-obesity-surgery.org/ accessed 2 May 2010.
4 One exception to this rises from a cluster of research articles suggesting that girls with 'absent' fathers are more likely to go through puberty early. Although divorce and single parenthood are difficult to control, one recent media report exhorted fathers to 'start seriously bonding with their daughters if they do not want them to turn into women at the age of 11' (James, 2009: 2).
5 Quantitative trait loci are identified through the statistical analysis of phenotypic and genotypic data with the goal of explaining the genetic basis of variation in complex traits (Miles and Wayne, 2008).
6 Gard and Wright (2005) critique the thrifty gene hypothesis – or what they refer as a 'just so' evolutionary story' (p. 111) on four key grounds: (1) that it is based on speculations about the behaviour of populations in distant history; (2) it presumes an extended period of stability in energy balance and constructs contemporary society in excessively generalised terms; (3) it generalises about our prehistoric ancestors based on arbitrarily acquired artefacts; and 4) it offers no basis for understanding the present since the hypothesis is more concerned with unstable food supply than energy expenditure.
7 Similarly, a recent study, published in *Nature Genetics*, suggests that contemporary trends in both early puberty and childhood obesity may 'involve a common metabolic response to the current nutritional environment' (Perry *et al.*, 2009: 3).
8 Although the National Institute for Health and Clinical Excellent (NICE) guidelines on the management of obesity approve surgery as an option in principle for those under 16, this is restricted to those who are deemed to have 'achieved or nearly achieved physiological maturity' (NICE, 2006: 43).
9 The term 'central precocious puberty' (CPP) refers to puberty that is dependent on the secretion of a particular hormone (gonadotropin-releasing hormone, GnRH) from the brain. It is usually idiopathic, although can be caused by forms of brain damage or dysfunction. Precocious puberty that is independent of GnRH (caused by sex hormones from other sources) is known as peripheral precocious puberty (PPP). Our focus here is on CPP.
10 See also articles published by the now defunct Intersex Society of North America on their archived website (http://www.isna.org).
11 See also Wardle *et al.* (2008) for a discussion of how genetic testing might relieve parents of responsibility for obesity in their children.
12 The standard forms of treatment of precocious puberty increase fat mass in girls' bodies and decrease lean mass and bone mass. There is no evidence, according to Carel and Leger (2008: 2372) that the treatment leads to obesity.

References

Anon., (2007), 'Girls with early puberty, older boyfriends at greater risk for drugs, sex, alcohol', *MedicalNewsToday.com*, (25 March, 2007), http://www.medicalnewstoday.com/articles/66073.php

Aphramor, L., (2005), 'Is a weight-centred health framework salutogenic? Some thoughts on unhinging certain dietary ideologies', *Social Theory and Health*, 3: 315–340.

Astrup, A., J.O. Hill and A. Rossner, (2004), 'The cause of obesity: are we barking up the wrong tree?' *Obesity Reviews*, 5: 125–127.

Banerjee, I. and P. Clayton, (2007), 'The genetic basis for the timing of human puberty', *Journal of Neuroendocrinology*, 19: 831–838.

Bray, G.A. and L.A. Tartaglia, (2000), 'Medicinal strategies in the treatment of obesity', *Nature*, 404(6): 672–677.

Carel, J-C. and J. Léger, (2008), Precocious Puberty, *The New England Journal of Medicine*, 358: 2366–2377.

Chagnon, Y.C., L. Perusse, S.J. Weisnagel, T. Rankinen and C. Bouchard, (2000), 'The human obesity gene map: the 1999 update', *Obesity Research*, 8(1): 89–117.

Cohen, S., (2002), *Folk Devils and Moral Panics* (3rd edn), London: Routledge.

Colls, R., (2007), 'Making sense of measurements: children's experiences of the BMI', seminar paper, presented at *Measuring Children*, 'Treating Fatness' seminar series, Centre for the Study of Women and Gender, University of Warwick (9 November, 2007).

Cooper, C., (1998), *Fat and Proud: the Politics of Size*, London: The Women's Press.

Creighton, S. and C. Minto, (2001), 'Editorials: Managing intersex: Most vaginal surgery in childhood should be deferred', *British Medical Journal*, 323: 1264–1265.

Cross Government Obesity Unit, Department of Health and Department of Children, Schools and Families, (2008), *Healthy Weight, Healthy Lives: A cross-Government Strategy for England*, London: HM Government.

DiVall, S.A. and S. Radovick, (2008), 'Pubertal development and menarche', *Annals of the New York Academy of Science*, 1135: 19–28.

Ebling, F.J.P., (2005), 'The neuroendocrine timing of puberty', *Reproduction*, 129: 675–683.

Ellis, B., (2004), 'Timing of pubertal maturation in girls: An integrated life history approach', *Psychological Bulletin*, 130(6): 920–958.

Evans, J., B. Evans and E. Rich, (2003), 'The only problem is, children will like their chips': education and the discursive production of ill-health', *Pedagogy, Culture and Society*, 11(2): 215–240.

Evans, J., E. Rich and R. Holroyd, (2004), 'Disordered eating and disordered schooling: what schools do to middle class girls', *British Journal of Sociology of Education*, 25(2): 123–142.

Evans, B., (2007), 'Doing more good than harm – every child matters? The absent presence of children's bodies in (anti-)obesity policy', seminar paper, presented at *Measuring Children*, 'Treating Fatness' seminar series, Centre for the Study of Women and Gender, University of Warwick (9 November, 2007).

Fausto-Sterling, A., (2004), 'Refashioning Race: DNA and the politics of health care', *Differences*, 15(5): 2–37.

Frayling, T.M., N.J. Timpson, M.N. Weedon, E. Zeggini, R.M. Freathy, C.M. Lindgren, J.R.B. Perry, K.S. Elliott, H. Lango, N.W. Rayner, B. Shields, L.W. Harries, J.C. Barrett, S. Ellard, C.J. Groves, B. Knight, A-M. Patch, A.R. Ness, S. Ebrahim, D.A. Lawlor, S.M. Ring, Y. Ben-Shlomo, M-R. Jarvelion, U. Sovio, A.J. Bennett, D. Malzer, L. Ferrucci, R.J.F. Loos, I. Barroso, N.J. Wareham, F. Karpe, K.R. Owen, L.R. Cardon, M. Walker, G.A. Hitman, C.N.A. Palmer, A.S.F. Doney, A.D. Morris, G.D. Smith, The Wellcome Trust Case Control Consortium, A.T. Hattersley and M.I. McCarthy, (2007), 'A Common Variant in the FTO Gene Is Associated with Body Mass Index and Predisposes to Childhood and Adult Obesity', *Science*, 316: 889–894.

Friedman, J.M., (2000), 'Obesity in the new millennium', *Nature*, 404: 632–634.

Friedman, J.M., (2004), 'Modern science versus the stigma of obesity', *Nature Medicine*, 10(6): 563–569.

Gard, M. and J. Wright, (2005), *The Obesity Epidemic: Science, Morality and Ideology*, London: Routledge.

Gianetti, E. and S. Seminara, (2008), 'Kisspeptin and KISS1R: a critical pathway in the reproductive system', *Reproduction*, 136: 295–301.

Graber, J.A., J. Brooks-Gunn and M.P. Warren, (1999), 'The vulnerable transition: puberty and the development of eating pathology and negative mood', *Women's Health Issues*, 9(2): 107–114.

Herman-Giddens, M.E., P.B. Kaplowitz and R. Wasserman, (2004), 'Navigating the recent articles on girls' puberty in pediatrics: What do we know and where do we go from here?' *Pediatrics*, 113: 911–917.

Hewitt, J.K., (1997), 'The genetics of obesity: what have genetic studies told us about the environment', *Behavior Genetics*, 27(4): 353–358.

Hill, J. and J.C. Peters, (1998), 'Environmental contributions to the obesity epidemic', *Science*, 280: 1371–1374.

James, P.T., R. Leach, E. Kalamara and M. Shayeghi, (2001), 'The Worldwide Obesity Epidemic', *Obesity Research*, 9(Supplement 4): 228S–233S.

James, O., (2009), 'Family under the microscope: Do absent fathers trigger early puberty in girls?' *The Guardian*, (March 28, 2009), http://www.guardian.co.uk/lifeandstyle/2009/mar/28/early-puberty-absent-fathers.

Keller, E., (2008), 'Nature and the natural', *BioSocieties*, 3: 117–124.

Kopelman, P.G., (2000), 'Obesity as a medical problem', *Nature*, 404: 635–643.

Latner, J.D. and A.J. Stunkard, (2003), 'Getting worse: the stigmatization of obese children', *Obesity Research*, 11(3): 452–456.

Lichtenbelt, W.D.V.M., J.W. Vanhommerig, N.M. Smulders, F.L. Drossaerts, G.J. Kemerink, N.D. Bouvy, P. Schrauwen and G.J.J. Teule, (2009), 'Cold-activated brown adipose tissue in healthy men', *New England Journal of Medicine*, 360(18): 1500–1508.

Loos, R.J.F. and C. Bouchard, (2003), 'Obesity – is it a genetic disorder?' *Journal of Internal Medicine*, 254: 401–425.

Mann, T., J. Tomiyama, E. Westling, A. Lew, B. Samuels, and J. Chatman, (2007), 'Medicare's search for effective obesity treatments', *American Psychologist*, 62(3): 220–233.

Marti, A., M.J. Moreno-Aliaga, J. Hebebrand and J.A. Martinez, (2004), 'Genes, lifestyles and obesity', *International Journal of Obesity*, 28: S29–S36.

Martin, N., (2008), 'Jamie Oliver apologises to mother he called "a scrubber"', *The Telegraph* (30 September, 2008) http://www.telegraph.co.uk/news/newstopics/celebritynews/3085793/Jamie-Oliver-apologises-to-mother-he-called-a-scrubber.html

McKie, R., (2005), 'Chemical kiss turns kids into adolescents', *The Observer*, (31 July, 2005) http://www.guardian.co.uk/society/2005/jul/31/health.medicineandhealth

McRae, F., (2007), 'Proof: there IS a gene that makes you fat', *Daily Mail* (16 April 2007), http://www.dailymail.co.uk/news/article-448160/Proof-There-IS-gene-makes-fat.html

Messager, S., (2005), Kisspeptin and its Receptor: New Gatekeepers of Puberty, *Journal of Neuroendocrinology*, 17(10): 687–688.

Miles, C.M. and M. Wayne, (2008), 'Quantitative Trait Locus (QTL) Analysis', *Nature Education*, 1(1).

Moshe, P. and L. Lazar, (2005), Precocious Puberty: Growth and Genetics Hormone, *Research in Pediatrics*, 64(Suppl. 2): 56–61.

Murray, S., (2005), '(Un/be)coming out? Rethinking fat politics', *Social Semiotics*, 15(2): 153–163.

Mul, D. and I.A. Hughes, (2008), 'The use of GnRH agonists in precocious puberty', *European Journal of Endocrinology*, 159: 3–8.

Neel, J.V., (1999), 'Diabetes mellitus: A "Thrifty" Genotype Rendered Detrimental by "Progress"?' *Bulletin of the World Health Organisation*, 77(8): 694–703.

NICE, (2006), *Obesity: the prevention, identification, assessment and management of overweight and obesity in adults and children. NICE Guideline. First draft for consultation, March 2006*, London: National Institute for Health and Clinical Excellence.

North, K.E. and L.J. Martin, (2009), 'The importance of gene environment interaction: implications for social scientists', *Sociological Methods and Research*, 37(2): 164–200.

Ogilvie, D. and N. Hamlet, (2005), 'Obesity: the elephant in the corner', *British Medical Journal*, 331: 1545–1548.

O'Sullivan, E., M. O'Sullivan and N. Mann, (2002), 'Precocious puberty: A parent's perspective', *Archives of Diseases of the Child*, 86: 320–321.

Palmert, M.R. and P.A. Boepple, (2001), 'Variation in the timing of puberty: clinical spectrum and genetic investigation', *Journal of Clinical Endocrinology and Metabolism*, 86(6): 2364–2368.

Parent, A-S., G. Teilman, A. Juul, N.E. Skakkebæk, J. Toppari and J-P. Bourguignon, (2003), 'The timing of normal puberty and the age limits of sexual precocity: variations around the world, secular trends, and changes after migration', *Endocrine Reviews*, 24(5): 668–693.

Perrin, A.J. and H. Lee, (2007), 'The undertheorized environment: sociological theory and the ontology of behavioural genetics', *Sociological Perspectives*, 50(2): 303–322.

Perusse, L. and C. Bouchard, (2000), 'Gene-diet interactions in obesity', *The American Journal of Clinical Nutrition*, 72: 1285–90S.

Perry, J.R.B., L. Stolk, N. Franceschini, K.L. Lunetta, Z. Guangju, P.F. McArdle, A.V. Smith, T. Aspelund, S. Bandinelli, E. Boerwinkle, L. Cherkas, G. Eiriksdottir, K. Estrada, L. Ferrucci,

A.R. Folsom, M. Garcia, V. Gudnason, A. Hofman, D. Karasik, D.P. Kiel, L.J. Launer, J. can Meurs, M.A. Nalls, F. Rivadeneira, A.R. Shuldiner, A. Singelton, N. Soranzo, T. Tanaka, J.A. Visser, M.N. Weedon, S.G. Wilson, V. Zhuang, E.A. Streeten, T.B. Harris, A. Murray, T.D. Spector, E.W. Demerath, A.G. Uitterlinden and J.M. Murabito, (2009), 'Meta-analysis of genome-wide association data identifies two loci influencing age at menarche', *Nature Genetics*, Advanced online publications, (May 17, 2009).

Phillip, M. and L. Lazar, (2005), 'Precocious puberty: growth and genetics', *Hormone Research*, 64(suppl 2): 56–61.

Posner, R.B., (2006), 'Early menarche: A review of research on trends in timing, racial differences, etiology and psychosocial consequences', *Sex Roles*, 54: 315–322.

Randerson, J. (2007), 'Obesity is not just gluttony – it may be in your genes', *Guardian*, (Friday 13 April 2007), http://www.guardian.co.uk/science/2007/apr/13/genetics.frontpagenews (accessed 31 March, 2009).

Rankinen, T., A. Zuberi, Y.C. Chagnon, S.J. Weisnagel, G. Argyropoulos, B. Walts, L. Perusse, C. Bouchard, (2006), 'The human obesity gene map: the 2005 update', *Obesity Research*, 14(4): 529–644.

Rayner, Jo-Anne, (2009), 'Cosmetic Endocrinology: (Re)constructing femininity in tall girls', Unpublished PhD Thesis, La Trobe University.

Roberts, M., (2005), 'Why puberty now begins at seven', *BBC News*, (May 15, 2005), http://news.bbc.co.uk/1/hi/health/4530743.stm

Roseweir, A.K. and R.P. Millar, (2009), 'The role of kisspeptin in the control of gonadotropin secretion', *Human Reproduction Update*, 15(2): 203–212.

Rosmond, R., (2004), 'Aetiology of obesity: a striving after wind?' *Obesity Reviews*, 5: 177–181.

Rossner, S., (2002), 'Obesity: the disease of the twenty-first century', *International Journal of Obesity*, 26 (Supplement 4): S2–S4.

Shell, E.R., (2002), *The Hungry Gene: the Science of Fat and the Future of Thin*, London: Atlantic Books.

Snyder, E.E., B. Walts, L. Perusse, Y.C. Chagnon, S.J. Weisnagel, T. Rankinen and C. Bouchard, (2004), 'The human obesity gene map: the 2003 update', *Obesity Research*, 12(3): 369–439.

Speakman, J.R., (2004), 'Obesity: the integrated roles of environment and genetics', *The Journal of Nutrition*, 134: 2090S–2105S.

Speakman, J.R., (2006), 'Thrifty genes for obesity and the metabolic syndrome – time to call off the search', *Diabetes and Vascular Disease Research*, 3(1): 7–11.

Spiegelman, B.M. and J.S. Flier, (2001), 'Obesity and the regulation of energy balance', *Cell*, 104(4): 531–543.

Stinson, K., (2001), *Women and Dieting Culture: Inside a Commercial Weight Loss Group*, New Brunswick: Rutgers University Press.

Throsby, K., (2009), 'The war on obesity as a moral project: weight loss drugs, obesity surgery and negotiating failure', *Science as Culture*, 18(2): 201–216.

Tremblay, A., L. Perusse and C. Bouchard, (2004), 'Energy balance and body-weight stability: impact of gene-environment interactions', *British Journal of Nutrition*, 92: S63–S66.

Wann, M., (1998), *Fat!So? Because You Don't Have to Apologise for your Size*, Berkeley, CA: Ten Speed Press.

Wardle, J., (2005), 'Understanding the aetiology of childhood obesity: implications for treatment', *Proceedings of the Nutrition Society*, 64: 73–79.

Wardle, J., S. Carnell, C.M.A. Haworth and R. Plomin, (2008), 'Evidence for a strong genetic influence on childhood adiposity despite the force of the obesogenic environment', *American Journal of Clinical Nutrition*, 87: 398–404.

Wehkalampi, K., K. Silventoinen, J. Kaprio, D.M. Dick, R.J. Rose, L. Pulkkinen and L. Dunkel, (2008), 'Genetic and environmental influences on pubertal timing assessed by height growth', *American Journal of Human Biology*, 20: 417–423.

Weinsier, R.L., G.R. Hunter, A.F. Heini, M.I. Goran and S.M. Sell, (1998), 'The Etiology of obesity: relative contribution of metabolic factors, diet and physical activity', *American Journal of Medicine*, 105: 154–150.

Part Four
Novelty and/in Nature?

Synthetic biology: constructing nature?

Jane Calvert

Introduction

Synthetic biology is a new scientific field which literally aspires to construct nature, by building living things 'from scratch'. Because of this approach, it challenges our ideas about what we should think of as 'natural'. An important aspect of how we understand 'natural' rests on what we oppose it to (see Keller, 2008a). In synthetic biology the main dichotomy is between the natural and the artificial. But other oppositions also become relevant: particularly those between the natural and the social, and the natural and the invented.

In what follows, I start with a brief description of synthetic biology and how it distinguishes itself from previous biotechnologies. I analyse the principles of an engineering approach to biology and show how these principles lead to aspirations amongst synthetic biologists to eliminate or reduce biological complexity in their synthetic creations. While some synthetic biologists want to 'improve' on nature (ie to make it easier to engineer), others want to replace certain natural phenomena with 'unnatural' alternatives. However, sceptics and critics argue that these blatant attempts to reduce complexity will not work, and that nature's contingency will not be successfully eliminated.

I then connect my discussion of synthetic biology to theoretical work by Paul Rabinow and Hans-Jörg Rheinberger on earlier biotechnologies, and discuss their analyses of the imposition of 'society' on 'nature' that we have seen in previous attempts to re-write and engineer biological systems. I go on to contextualize our understanding of synthetic biology and its construction of nature by considering model organisms and intellectual property, both of which attempt to impose uniform properties of natural entities, and which have the potential to move their objects out of the realm of the natural and into the realm of the artificial. In the conclusions I show how pressures for engineerability, commodification and standardization are all pulling in the same direction: towards a reconstruction of nature which is instrumentalizable and utilizable for our purposes. These pressures could have profound consequences for the kinds of living things that are brought into the world by synthetic biology.

This chapter draws on two years of research on the emerging field of synthetic biology. It is based on my experiences in being the social science member of a UK working party on synthetic biology, a founder member of a UK cross-university synthetic biology network, and my involvement in a synthetic biology 'sandpit'(which involved developing research proposals with synthetic biologists), as well as participation in numerous workshops, conferences and meetings in synthetic biology, from the largest conference in the field so far (Synthetic Biology 4.0 in Hong Kong, 2008), to meetings directed to young researchers and postgraduates (eg BioSysBio in Cambridge, 2009), to small discussion meetings internal to my own university and elsewhere. Detailed records have been kept throughout of my observations, conversations, interactions and reflections, so the empirical basis for this chapter is best described as a multi-sited ethnography (Marcus, 1995). Synthetic biologists are only quoted here by name when I refer to a comment made in a public meeting; otherwise they are quoted by anonymized code name. I also draw on the burgeoning scientific literature in synthetic biology.

Synthetic biology

It is hard to provide a precise definition of any new and emerging scientific field, but defining synthetic biology is particularly difficult because it incorporates a number of disparate research activities under its banner. Broadly speaking, synthetic biology can be described as a field which aims to construct living systems *de novo*. The new kinds of biological entity it produces can be construed as being extremely unnatural and 'synthetic' – as the term 'synthetic biology' implies. Although 'synthesis' also has the meaning of synthesizing or putting together (ie the opposite of 'analysis'), it is the meaning of synthetic as 'artificial' which is the one that many commentators draw upon when discussing synthetic biology. In fact, attempts have been made to avoid the word 'synthetic' by naming the field 'constructive biology' or 'intentional biology' (Carlson, 2006), but these names have not become widely adopted.

At the moment there is no consensus on the definition of the field. One of the most common definitions is a two-pronged one, which includes the construction of completely novel biological entities, and the re-design of already existing ones. For example, a group of leading scientists in the field defines synthetic biology as 'the design and construction of new biological parts, devices, and systems and the re-design of existing, natural biological systems for useful purposes'.[1]

Despite the lack of consensus, it is helpful to divide the disparate field of synthetic biology into three broad approaches (see O'Malley *et al.*, 2008). The first approach, which is my main focus here, involves the building of interchangeable biological parts and devices called 'BioBricks' (see Endy, 2005). The second approach covers work at the level of whole genomes, including both the synthesis of viral genomes from scratch (eg Cello *et al.*, 2002) and attempts to strip away excess DNA from existing bacterial genomes, with the objective of producing the minimal genome necessary for life (Glass *et al.*, 2006). The third approach centres on the creation of 'protocells' from simple components such as lipid vesicles (see Deamer, 2005; Luisi *et al.*, 2006).

An overriding principle, which governs all three approaches, is the attempt to engineer life, to such an extent that synthetic biology is called 'the engineer's approach to biology' (Breithaupt, 2006). For example, the 'BioBricks' school, which is the dominant approach to synthetic biology, draws on the engineering principles of standardization, decoupling and abstraction with the objective of developing biological components which are interchangeable, functionally discrete and capable of being combined in a modular fashion, along the lines of 'plug and play' (Isaacs and Collins, 2005; Brent, 2004). These synthetic biologists hope that they will succeed in making biology into an engineering discipline where genetic 'engineering' failed (Breithaupt, 2006; Endy, 2005).

For this reason, synthetic biologists commonly distinguish their work from genetic engineering. There are two ways in which they do this. One is in terms of the methods used. It is argued that since synthetic biology uses standardized parts and follows a formalized design process, the tools and intellectual approach of engineering are adopted in a way which makes synthetic biology more authentically like engineering. Another distinction is in terms of the 'sophistication' of the work. In genetic engineering one gene at a time is modified or added, whereas in synthetic biology a whole specialized metabolic unit can be constructed. This means that in synthetic biology we see engineering at the systems level, rather than at the level of the individual components. It should also be noted that another reason why synthetic biologists may be keen to distinguish their field from genetic engineering is because of some of the perceived negative social implications of genetic engineering.

The reason why synthetic biologists want to make biology into an engineering discipline is because they want to replicate the successes of engineering in a biological context. Synthetic biologists draw inspiration from the technological achievements of other branches of engineering, such as aircraft and computers, and conclude that it is 'economically and socially important that we improve the efficiency, reliability and predictability of our biological designs' (Arkin, 2008, p. 774). Parallels are often drawn between today's synthetic biology and the early days of the nascent computer industry, with the intended implication that the technological revolution that synthetic biology brings will be as important as the revolution in ICTs brought about by electrical engineering (Barrett *et al.*, 2006; NEST, 2005).

As with other types of engineering, the range of applications of synthetic biology is potentially very broad, although the field is currently at an early stage of development. Examples of potential applications include microbial biofuel production (Keasling and Chou, 2008), bioremediation, biosensors, new drug development pathways and synthetic vaccines (Royal Academy of Engineering, 2009). Most notably, synthetic biology has successfully been used to produce a precursor for an antimalarial drug (Martin *et al.*, 2003).

The reduction of complexity

While other emerging areas of biology (such as systems biology), embrace the complexity of biological systems, a key feature of the engineering approach to synthetic biology is the attempt to reduce biological complexity. Programmatic statements along these lines are common. For example, Heinemann and Panke (2006) say: 'As the complexity of existing biological systems is the major problem in implementing synthetic biology's engineering vision, it is desirable to reduce this complexity' (p. 2793). Leading synthetic biologist Tom Knight also sees this aspiration to get rid of complexity as an integral part of the engineers' approach in maintaining that 'an alternative to understanding complexity is to get rid of it' (quoted in Ball, 2004). One of the founders of synthetic biology, George Church, adopts a similar line: 'You focus on parts of the science that you do understand and clean out the parts that you don't understand' (quoted in Breithaupt, 2006: 22–3).

The approach we see here towards the complexity of nature is that it is better dispensed with, and that we will be more successful if we attempt to construct living systems from scratch without the unnecessary detritus that they have accumulated over evolutionary time. An example of this approach towards complexity is seen in the key paper 'Refactoring Bacteriophage T7' (Chan *et al.*, 2005). The word 'refactoring' here is borrowed directly from computer software, and it means improving computer software by rationalising it and 'cleaning it up'.[2]

For some this reduction of complexity is not merely an instrumental aim, but is based on a faith that synthetic biology will ultimately lead to 'the elucidation of the underlying simplicity' of nature (Palsson, 2000: 1149). The pervasive idea of the simplicity of nature, and its connection to truth, is found throughout the history of biology (often in tension with ideas about its complexity – see Keller, 2008b), and perhaps the most striking example of this is the iconic image of the double helix, famously described by Crick and Watson in 1953. Synthetic biologists draw inspiration from discoveries such as these and hope that the complexity of biological systems will be shown to be an eliminable accident of historical accumulations over evolutionary time (Balaram, 2003).

Modularity in synthetic biology

A key requirement for the reduction of complexity is the assumption of modularity. Modularity is not a straightforward concept, but in engineering terms a module is defined as 'a functional unit that is capable of maintaining its intrinsic properties irrespective of what it is connected to' (Sauro, 2008: 1). Modular entities are very important in engineering because they can be extracted from one part of a system and inserted in another with no change in their function (Sauro, 2008). Modularity is crucial to the BioBricks school of synthetic biology where there is an on-line 'Registry of Standard Biological Parts'[3] containing modular BioBricks that, in theory, can be connected to one another in a manner similar to pieces of lego (an analogy explicitly exploited by the Registry).

The issue of modularity is very interesting in the context of the discussion of nature. The key question here is whether biological systems are actually comprised of functional modules, or if they are simply best understood as such by the engineering approaches that are adopted in synthetic biology. There is no consensus on this issue. Arkin and Fletcher (2006) say that 'The key observation that biological systems exhibit some degree of modularity underlies the current belief that useful and 'engineerable' design principles exist' (p. 2). And there are examples of entities that, at first glance, do appear to be modular and discrete, such as cells (Kitano, 2002), ribosomes (Hartwell *et al.*, 1999) and antibiotics (Synthetic biologist 5, November 15, 2008).

But there are others who see the idea of modularity as one that is imposed by the scientists and engineers involved. For example, even the proponents of abstraction hierarchies (the division of synthetic biology into parts, devices and systems that underlies the BioBricks approach) admit that these hierarchies are 'a human invention designed to assist people in engineering very complex systems by ignoring unnecessary details'.[4] We cannot necessarily deduce from this statement that these synthetic biologists think that modules are impositions of the engineer upon the biological substrate, but we can see a recognition that some aspects of synthetic biology are imposed on nature, rather than found within it. Other synthetic biologists argue that modularity and standardization in nature are myths, because biological function is always context dependent (Synthetic biologist 5, November 15, 2008).

Most synthetic biologists, however, do maintain that the modularity they aspire to find and build is present in biology, but even when agreement is reached on this issue, there is still a great deal of discussion over what constitutes a module, and whether this varies depending on the disciplinary perspective one adopts. The problem is that 'modules mean different things to different specialists: the module of the developmental biologist is not the same as the module of the molecular geneticist' (Morange, 2009: S50). Furthermore, biological entities have evolved to survive and not to be conveniently structured so that scientists can understand them (Keller, 2008b). This means that what a cell regards as a functional module may be different from what a synthetic biologist regards as one (Oliver, 2008). It may be that we are carving out modules in nature to fit our desires for biological understanding. Keller (2008b) warns that even if an assumption of modularity is helpful in the sense of enabling us to gain an understanding of biology (in an epistemological sense), we must be careful that we are not confusing epistemology with ontology.

There is also the point that all scientific knowledge must involve simplification and decontextualization to an extent, in order to better focus on the subject of interest. As Barnes and Dupré (2008) note, in all attempts to gain knowledge some kind of reduction of complexity is perhaps inevitable: even perception involves reduction and simplification in order to prioritize the perceptual inputs that are most important.

But synthetic biology is particularly interesting in respect to modularity, because even if biology is not modular, perhaps synthetic biologists can make it so. The engineers' somewhat bullish approach to eliminating annoying noise and crosstalk may result in an organism that is actually more modular than 'natural' organisms are. (This raises interesting questions about what such organisms might look like, and animals made of lego – one of the favourite analogies of the BioBricks school – spring to mind as a visual aid).

Attempts to break biological systems down into modular components not only makes biological complexity easier to deal with, but also makes these components more similar to software code which is modular, standardized and re-useable. One advantage of modularity is that several different researchers can work on different parts simultaneously, meaning that the field can develop faster. In this way, modularity is well-suited to open source principles, which require discrete elements to be open to continual modification and improvement by a large community of users (see Calvert, 2008). Many synthetic biologists are ideologically committed to open source, to such an extent that the aspiration to make their work open source is a guiding principle of the field.

Credit: Satoshi Kamayashi

Improving on nature

The idea of making living things more modular than they are 'naturally' gives rise to the potential of synthetic biology to improve upon nature. The idea underlying much synthetic biology is that making biological systems from scratch will make them better than they are in nature. Vitor Martins dos Santos (2008) describes synthetic biology along these lines as following three steps: understand the wiring, 'reprogram' and simplify. And the tag line which appears at the bottom of every page of the website of syntheticbiology.org is 'making life *better*, one part at a time'.[5]

It is pertinent here to ask what understanding of 'better' is being adopted in this tag line. From the perspective of a synthetic biologist, making life 'better' is making it easier to engineer. Here there is the potential for the imposition of engineering principles on the form of living things, and arguably this is something that we are already starting to see. One of the main research objectives in synthetic biology has been to build analogues of engineering, such as oscillators and logic gates, in biological systems.[6] We see here the hope that synthetic biology will mimic the successes of electrical engineering, and it is telling that Hartwell *et al.* (1999) proclaim that 'The next generation of [biology] students should learn how to look for amplifiers and logic circuits' (p. C52). These types of synthetic biological creations have been constructed on the basis of the assumptions that biological systems are modular and that feedback loops are important in explaining their operation. However, it has not been proven that these assumptions apply to 'natural' biological systems (Loettgers, 2007).

Some of the more sceptical commentators see this approach as an imposition of the engineering mentality on natural systems, arguing that oscillators are rare in actual living systems (Synthetic biologist 6, 25th January 2008). Pottage and Sherman also make the point that 'the image of synthetic biology as an exercise in 'engineering' building blocks and programmable logic gates synthesized from inanimate materials, extends the mechanical and instrumental vision of nature into the deep texture of life' (2007: 545).

A good example of the way in which engineering principles are imposed by synthetic biology is found in one of the most important papers in the field which describes the building of a 'repressilator' (Elowitz and Leibler, 2000), a genetic regulatory network that exhibits stable oscillation (although from an engineering point of view it is noisy and inflexible, Kitney, 2008). What is interesting about the construction of the repressilator is that the aim of the work was not to reproduce experimental findings as accurately as possible. Instead, the researchers 'designed a network which would allow them to study not *natural* network designs but *possible* network design' (Loettgers, 2007: 141, emphasis added), so that they could learn more about how to develop alternative biological structures.

Although these synthetic networks may be functioning primarily as heuristic tools at the moment, in the process of doing synthetic biology these heuristic tools become material constructions. Again, there is the potential for the blurring of ontology and epistemology, because the reshaping of nature in synthetic biology is tied up with scientists' own epistemic practices.

Another rather audacious example of how epistemic practices can come to influence biological materiality is given by synthetic biologists who say that if they discover that their models of biological phenomena do not work, then they will simply engineer the biological parts to fit the model better (Synthetic biologist 8, 16th January 2009), rather than changing the model to make it a more accurate representation of the biological system, which is how biological research would have proceeded in the past.

A different kind of example of the imposition of engineering principles onto the natural world is the attempt to create a minimal 'chassis' into which biological functions can be slotted. The most cited example is the Venter Institute's work, where scientists are attempting to strip away excess DNA from an existing microbial genome until they are left with the minimal genome which is necessary for life (Glass *et al.*, 2006). Although the chassis school of synthetic biology is distinct from the BioBricks school, the aspiration is that the two will coincide in the future because the chassis will provide the substrate into which new BioBricks can be embedded.

It is worth reflecting on the metaphor of a 'chassis' here, and thinking about what work is being done with this familiar mechanical analogy: can we really think of cells in these terms and consider slotting in different functional modules of DNA? The idea that it is possible to reduce 'life' down to its bare minimum, and then insert new functional parts into this minimum is a clear demonstration of an instrumentalist approach.

Changing nature

As well as attempts to make nature 'better' by stripping away unnecessary complexity, there are also attempts to make a biology that is different from that which is found in nature altogether. One of the objectives driving this work is to push the boundaries of natural systems and in this way learn more about them. Examples of this approach are Chin's work, which involves producing amino acids from 4 instead of the normal 3 codons (see Chin, 2008). An interest in the limitations of natural systems also drives research which attempts to develop new nucleic acids beyond the familiar As, Ts, Cs and Gs (Pollack, 2001). One motivation for this creation of 'unnatural' biological systems is to make modularity more successful. The idea here is that 'orthogonal' parts could be designed, which perform the same function as their natural counterparts but which, because they are not naturally found in the cell, do not interfere with the existing cellular context and thus are more likely to be easily separable and manipulable (Weiss, 2008).

This idea of orthogonal, alternative versions of natural systems is interestingly, and somewhat counter-intuitively, also used to argue for the safety of synthetic biological constructs. For example, it has been suggested that synthetic organisms could be made to be dependent on nutrients that are not found in nature (De Vriend, 2006), or that they could have built-in safety features such as 'fail-fast' mechanisms (Endy, 2005). One of the arguments made in favour of alternative biological constructs is that we would not see 'interbreeding' between artificial and natural organisms (Lancet, 2008). Here the assumption is that making synthetic organisms *less* natural will make them less risky, because they will be more easily separable from the natural world.

Others point out that pushing the boundaries of natural systems in this way could lead to new and completely unexpected problems. Putting genes together that were previously separate could lead to the creation of new organisms that have unpredictable emergent properties (Tucker and Zilinskas, 2006), making the risks of accidental release very difficult to assess in advance (De Vriend, 2006).

Limitations to the reduction of complexity

The attempt to reduce complexity, which underlies much of this work, may not be achievable at all, however. The key question here is: 'Can the basic biological, evolutionary, non-linear aspects of living systems be engineered out?' (Rabinow, in Lentzos *et al.*, 2008: 315). Synthetic biologists are aware of these difficulties. For example, one says that the non-linearity of living systems makes them different from conventional engineering systems (Synthetic biologist 7, 5th Feb 2008), and Pam Silver describes a living organism as 'A long series of kludges . . . not necessarily a well-oiled machine' (Silver, 2008). The philosopher William Wimsatt (2007) colourfully elaborates by saying we should think of 'Nature as a reconditioned parts dealer and crafty backwoods mechanic, constantly fixing and redesigning old machines and fashioning new ones out of whatever comes easily to hand' (p. 10).

Synthetic biologists, such as those in Ron Weiss' group in Princeton, advise that 'it may be prudent to treat some biological uncertainties as fundamental properties of individual cell behavior' (Andrianantoandro *et al.*, 2006: 13). They argue for a 'middle way'; acknowledging the complexity of biological systems but simultaneously not assuming that every single biological part is going to be a one-off. This leads Weiss to the conclusion that in synthetic biology it becomes important to think hard about 'what makes the biological substrate not like the other substrates we usually deal with' (Weiss, 2008).

Others are more pessimistic and argue that we will see synthetic biology continually eluded in its quest to isolate the properties of living systems. The concern here is that by attempting to eliminate complexity and contingency, synthetic biologists might end up losing sight of the emergent properties that define living systems, which are themselves historical accumulations, being the result of billions of years of evolution (Balaram, 2003 and Dupré, forthcoming 2010). If this is correct we should not be optimistic about the attempts to construct life from modular and substitutable parts, which, in some guises, is the guiding objective of synthetic biology.

However, some commentators argue that those who point to the limitations of applying engineering principles to living systems are suffering from 'urea syndrome', ie they are assuming that there is something special and irreducible about living things and that it is by definition impossible to recreate them using scientific methods. This was the argument made about organic compounds before the successful synthesis of urea in 1828, which sent shockwaves through the scientific community of the time (Lazebnik, 2002).

Nature modelled on culture

The previous discussion has shown how synthetic biology, like other biotechnologies that preceded it, works by 'extending the reach of human manufactures into the texture of life itself' (Pottage, 2007: 324). In this way, the attempts to manipulate and control nature that we see in synthetic biology are not new, and there are important continuities between synthetic biology and previous biological research. As Franklin (2006) says: 'The biologization of human values is in some ways as old as horticulture, when human preferences began to be nudged into seedlings, and mutated corn began to be selected for its ears' (p. 179), but (to paraphrase her) this does not mean that synthetic biology does not have its own specificity. We could argue that what we see in synthetic biology is an extension of the potential of biotechnology, with improved capabilities for actually manipulating nature. It is not just that nature is being tamed, as it was with previous biologies. In synthetic biology nature is being explicitly remade.

The remodelling of the natural on the basis of an analogy with engineering that we see in synthetic biology is very similar to Rabinow's (1999[1992]) idea of 'biosociality', where, he says, 'nature will be modeled on culture' (p. 411). He calls this 'biosociality' because he reverses 'socio-biology', where culture is 'constructed on the basis of a metaphor of nature' (p. 411). This leads to an inversion of the temporal order in which we normally think about nature and culture, as Franklin (2000) explains: 'culture becomes the model for nature instead of being 'after nature', as if a kind of successor project' (p. 194–5). A consequence of this is that 'biodiversity becomes a product of design choices, and industrial and political imperatives . . . rather than evolutionary pressures' (Allenby, 2006).

Rabinow (1999 [1992]) anticipates the developments in synthetic biology by saying that 'Nature will be known and remade through technique and will finally become artificial' (p. 411). His work was written at the time of the sequencing of the human genome, and he predicts that what will be most interesting about the human genome is that it 'will be known in such a way that it can be *changed*' (p. 408 emphasis added). The modification of nature has perhaps always been the objective of the biotechnological enterprise. But we could argue that this remaking, or re-writing, has only really become properly possible with synthetic biology.

Rheinberger (2000) traces the potential for changing nature back to the development of recombinant DNA techniques. He describes this development by saying that in early molecular biology there was 'an extracellular representation of intracellular configurations' (p. 19). In other words, the scientific objective was merely to represent what was going on inside the cell. But with the advent of recombinant DNA 'a radical change of perspective ensued. The momentum of gene technology is based on the prospects of an intracellular representation of extracellular projects – the potential 'rewriting' of life' (p. 19). In this way ideas about what nature should be (ie extracellular, 'cultural' projects), could come to influence the intracellular environment. Rheinberger says that this potential for 'rewriting' makes molecular engineering 'substantially different from traditional intervention in the life sciences and medicine' because it involves '*re-programming* metabolic actions, not just interfering with them' (p. 25). Rheinberger sees a technological potential here for intervening in a real sense, a potential which is being exploited by synthetic biology. This leads Rheinberger to a very similar conclusion to Rabinow: 'the very essence of our being social is not to supersede, but to alter our natural, that is, in the present context, our genetic condition. We come to realise that the *natural* condition of our genetic makeup might turn into a *social* construct, with the result that the distinction between the 'natural' and the 'social' no longer makes good sense' (p. 29).

Rheinberger makes heavy use of the metaphors of writing and programming, claiming that 'Within a timespan of less than twenty years, molecular geneticists have learned not only to understand the language of genes in principle, but to spell it' (p. 24). Both these metaphors are ubiquitous in synthetic biology. For example, leaders of the field regularly talk about being able to 'write DNA' (Endy, 2004). For this metaphor to have clout, DNA is assumed to take the form of information, which can be read (with the techniques of gene sequencing), and now written (using the techniques of synthesis) (Villalobos *et al.*, 2006).[7] Many synthetic biology papers also talk about 're-programming' totally uncritically (Heinemann and Panke, 2006; Dueber *et al.*, 2003; Gallivan, 2007). To give an example, Gallivan (2007) says that 'One of the main goals of synthetic biology is to reprogram organisms to autonomously perform complex tasks' (p. 612). Slightly more reflection is found in a *Science* article, which at least recognizes that there is a gap between the organism and the computer programme in saying that 'Synthetic biologists eventually aim to make bacteria into tiny programmable computers.' (Ferber, 2004: 160).

Work by Rabinow and Rheinberger shows how culture and nature can be inverted, with 'cultural' frameworks (such as engineering and computation) having profound effects on the ways in which we construct living things. I hope to have demonstrated that these theoretical discussions help elucidate and enrich our understanding of synthetic biology.

Model organisms and intellectual property

Work in other areas can also contribute to an analysis of synthetic biology, and in this section I briefly discuss the attempts to standardize and control nature that we see in model organisms and intellectual property.

The modular biological parts that synthetic biologists are attempting to create have many similarities to the model organisms that have been standardized and homogenized so that they can be used reliably and predictably in laboratory work (Kohler, 1999). For example, a key objective behind the development of the model worm *C. elegans* was 'reducing complex data to a conceptually and experimentally tractable system' (Ankeny, 2000: S270). This is very reminiscent of the attempts to reduce complexity that we find in synthetic biology. In the case of the experimental mouse, the aim was to create mice which were so similar to each other that they were exchangeable, which is what we currently see in the development of modular BioBricks. We even see the use of the 'plug' metaphor in the mouse case:

> standardization is about what happens when one 'plugs in' a purified, specialized mouse into a research process (experiment, breeding program). If one such mouse is *substitutable* for another, then the mouse meets a standard of purity (Griesemer and Gerson, 2006: S366, italics in original).

The stabilization that we find in model organisms, which allows them to be useful laboratory tools, is also found in intellectual property, this time so that living things can be easily exchanged in the market place (see Calvert, 2008). For example, an argument made in favour of patenting microorganisms is that they 'are formed in such large numbers that any measurable quantity will possess *uniform properties and characteristics*.' (Judge Bastarache, Canada Supreme Court, 2002, in Dutfield, 2008: 3, emphasis added). If living organisms are uniform then they can be more easily treated as commodities.

A similar argument was made in the 1930 plant patent act – legislation which first allowed the patenting of new plant varieties in the United States. It was maintained that since asexually reproduced plants did not vary, they should be considered patentable (Pottage and Sherman, 2007). Furthermore, the reproduction of a new plant of this kind could not be accomplished by nature, but had to involve human intervention. In this way the role of the breeder 'was to normalise the abnormal, to stabilise and standardize nature's deviants, mutations and aberrations' (Pottage and Sherman, 2007: 559). With this stabilization and standardization work, the buyer would know that they were getting a reliable product. This emphasis on standardization and homogenization, and the attempt to eliminate aberrations is again very reminiscent of synthetic biology. The catalogues that plant breeders produced so that their wares could be bought as replicable copies (Pottage 2009) bear strong similarities to the 'catalog of parts and devices' that can be found on the BioBricks website.[8]

We see attempts to standardize in both model organisms and in intellectual property, but the two areas are linked in another way because model organism communities often share views about ownership. For example, the *Drosophila* community adopted strong norms of 'sharing and free exchange' (Kohler, 1999: 345), and *Drosophila* researchers are known for being particularly cooperative, even today. Similarly, the scientist who first produced the standardized experimental mouse did not aspire to profit from it, instead 'He favored traditional, cooperative exchange of materials as a service to the community of researchers' (Griesemer and Gerson, 2006: 366). The *C. elegans* community is also 'often celebrated as a model of scientific cooperation' (Ankeny, 2000: S262). It seems as if model organism communities have accompanying norms which favour the sharing of their standardized organisms. Perhaps this sharing is necessary in a context where scientists are attempting to encourage buy-in to a particular standard.

We saw above how an open source approach was favoured by many synthetic biologists, and that modularity could be seen as facilitating such an approach. This is another demonstration of the similarities between synthetic biology and model organism communities, and in this light, it is interesting how at the closing session of the Synthetic Biology 4.0 conference in Hong Kong in 2008, explicit parallels were drawn between the synthetic biology community and the *C. elegans* community.

Another particularly important point about the model organism communities was how in the historical cases described above, the model organism, the social arrangements and the intellectual property norms were all produced together (Kohler, 1999). In synthetic biology we also see explicit attempts among certain leaders of the field to build a community that shares a certain set of norms about open source (and also about biosafety), at the same time as we see the building of the BioBricks and the chassis that are the material building blocks of synthetic biology. This is a clear example of how natural and social orders are co-constructed (Jasanoff, 2006). And in this example, as in many others in this volume, 'it is abundantly clear that technoscience and its artefacts are central to remaking society and nature simultaneously' (Braun and Castree, 1998: 29).

The open-source strand of synthetic biology is not the only one, however. There are also synthetic biologists who are keen to gain proprietary ownership over their synthetic creations.

As mentioned above, the 'unnaturalness' of synthetic organisms is used to argue for their safety, and their unnaturalness can also be useful in a patenting context, because it can be used to argue for their inventiveness. This is because the 'product of nature doctrine' holds that if something already exists in nature then it is not patentable, because it is not a novel invention. This doctrine can be traced back to the landmark Diamond versus Chakrabarty decision on a patent on a modified microorganism, where the court concluded that something could only be patented if it was 'a nonnaturally occurring manufacture or composition of matter – a product of human ingenuity' (447 US § 303 [1980], p. 309 in Conley and Makowski, 2004). Since synthetic biology aims to de-complexify and improve on natural biological systems, its creations are clearly different from what is found in nature, so an argument can be made that they are human inventions, and that they deserve the reward of a patent.

Some synthetic biologists even purposely increase the unnaturalness of their creations, and make them more proprietary by marking them. Most notably, Craig Venter has 'watermarked' his name and the names of his collaborators into the code of his minimized bacterial genome, by inserting codons that produce proteins which correspond to the appropriate letters of the alphabet (Highfield, 2008). This is a good example of a way in which synthetic biology plays with the boundary between the natural and the artificial, in the context of intellectual property.

There is a broader point here. Some commentators see the whole of intellectual property law as drawing on an entrenched distinction between the natural and the artificial. When something becomes intellectual property it is moved out of the realm of the natural into the realm of the artificial; it becomes an artefact (Biagioli, 2007). But there are problems with presupposing that there is such a divide between the natural and the artificial, because it assumes that there is a stable concept of nature 'as such' – some kind of 'a pre-given substrate'

(Pottage, 1998). As we saw above, the arguments made by Rabinow and Rheinberger showed that culture can become a model for nature, making it very difficult to decide where culture ends and nature begins. Franklin points to exactly this problem when she says that

> The twentieth-century transformation of life itself has had the consequences that the grounding or foundational function of nature as a limit or force in itself has become problematic and lost its axiomatic, *a priori*, value as a referent or authority, becoming instead a receding horizon (Franklin, 2000: 190).

Conclusions

This chapter has examined the ways in which synthetic biology is constructing nature. It is literally constructing nature by building new types of biological entity, but it is also constructing a new understanding of nature by challenging our notions of what is 'natural'. For example, we have seen how some synthetic biologists see themselves as improving on nature, while others emphasize the unnaturalness of their creations, to fit with patent demands or to produce orthogonal systems that do not interfere with existing biological contexts. However, although 'synthetic' is sometimes used as a synonym for 'artificial', it is unlikely that we will happily allocate all synthetic biological creations to the realm of the artificial. The standardized, modular, decomplexified creations of synthetic biology will inevitably start to infect our understandings of what is 'natural', which, as we have seen, is itself a 'receding horizon' defined primarily in terms of what it is opposed to.

I have argued that the objective behind the (re)construction of nature in synthetic biology is instrumental: the aim is to 'improve' on nature in order to make it easier to engineer. A key aspect of this improvement is the reduction of complexity, because 'the more dramatically researchers can reduce the complexity of biological organisms, the better they can turn these organisms into instrumentalizable media' (Pottage, 2007: 330). These tendencies may be a symptom of the 'instrumentalist epistemology of modern scientific culture overall' (Wynne, 2005: 77), which Wynne sees as driving research towards prediction and control and the potential for exploitation.

I have shown how synthetic biology is not exceptional in its attempts to instrumentalize nature, but that there are a confluence of different factors which all push in the same direction, and which include engineering, modularity, scientific practices, model organisms, standardization, exchangeability, intellectual property, and even open source. But we have also seen how the desire for prediction and control may be thwarted, and how there is a tension between the standardization which synthetic biologists attempt to impose on the biological substrate, and nature's apparent unruliness. Even in an 'age of biological control' (Wilmut *et al.*, 2000), we see the non-cooperation of nature. A key objective of synthetic biology is to overcome this recalcitrance, which is why it is common to hear leading synthetic biologists talk of 'mastering nature' (Silver,

2009), and of wanting to 'manipulate biology to give us the kind of things we want and not accept what nature gave us' (Keasling, 2009). It will be fascinating to see how this tension will resolve itself in the context of synthetic biology, and what limitations the exuberance of nature will impose on the scientists and engineers' desires for control.

I will end by suggesting that the example of synthetic biology can give us some very interesting insights into how 'the world 'outside' the laboratory comes to mirror the world inside' (Braun and Castree, 1998: 27). For example, we have seen how the epistemic ideal of modularity is imposed on the materiality of living things, and how synthetic biologists prefer to change nature than change their models, if they discover their models do not work as predicted. This goes further than Hacking's (1992) observation that technoscience enacts worlds that are fit for its methods, because the powers of intervention in synthetic biology are potentially much greater than those of previous biotechnologies. In synthetic biology we are seeing a reconstruction of nature that is utilizable and instrumentalizable, and is a product of epistemic ideals and design choices. This reconstruction may have profound consequences for the types of living things that are brought into the world in the future.

Notes

1 www.syntheticbiology.org accessed 30 April 2010.
2 http://www.refactoring.com/ accessed 30 April 2010.
3 http://partsregistry.org/Main_Page accessed 30 April 2010.
4 http://www.openwetware.org/wiki/Synthetic_Biology:Abstraction_hierarchy accessed 30 April 2010.
5 www.syntheticbiology.org (emphasis added) accessed 30 April 2010.
6 Oscillators and logic gates are common components of electronic circuits. An oscillator produces a repetitive electronic signal and a logic gate switches the flow of electricity 'on' or 'off' depending on the input. accessed 30 April 2010.
7 For criticism of this understanding of DNA as informational see Barnes and Dupré (2008). accessed 30 April 2010.
8 http://partsregistry.org/Main_Page accessed 30 April 2010.

References

Allenby, B., (2006), 'Biology as cultural artifact', http://www.greenbiz.com/blog/2006/10/01/biology-cultural-artifact (accessed 30 May 2009).

Andrianantoandro, E., S. Basu, D.K. Karig and R. Weiss, (2006), 'Synthetic biology: new engineering rules for an emerging discipline', *Molecular Systems Biology*, doi: 10.1038/ msb4100073.

Ankeny, R.A., (2000), 'Fashioning Descriptive Models in Biology: Of Worms and Wiring Diagrams', *Philosophy of Science*, Vol. 67, Supplement. Proceedings of the 1998 Biennial Meetings of the Philosophy of Science Association. Part II: Symposia Papers, Sep., 2000: S260–S272.

Arkin, A., (2008), 'Setting the standard in synthetic biology', *Nature Biotechnology*, 26, 7: 771–774.

Arkin, A.P. and D.A. Fletcher, (2006), 'Fast, cheap and somewhat in control', *Genome Biol*, 7: 114.

Balaram, P., (2003), 'Synthesising life', *Current Science*, Vol. 85, no. 11: 1509–1510.

Ball, P., (2004), 'Starting from Scratch', *Nature*, 7th October, 2004, 431: 624–626.
Barnes, S.B. and J.A. Dupré, (2008), *Genomes and What to Make of Them*, Chicago, IL: University of Chicago Press.
Barrett, C.L., T.Y. Kim, H.U. Kim, B.Ø. Palsson and S.Y. Lee, (2006), 'Systems biology as a foundation for genome-scale synthetic biology', *Current Opinion in Biotechnology*, 17(5): 1–5.
Biagioli, M., (2007), 'Denaturalizing the public domain: How to use science studies to rethink IP', Talk at the University of Edinburgh, 10th December 2007.
Braun, B. and N. Castree, (1998) *Remaking reality: nature at the millennium*, London and New York: Routledge.
Breithaupt, H., (2006), 'The Engineer's Approach to Biology', *EMBO Reports*, 7(1): 21–24.
Brent, R., (2004), 'A partnership between biology and engineering', *Nature Biotechnology*, 22(10): 1211–1214.
Calvert, J., (2008), 'The commodification of emergence: systems biology, synthetic biology and intellectual property', *BioSocieties*, 3(4): 385–400.
Carlson, R., (2006), 'Synthetic biology 2.0, Part IV: What's in a name?' http://synthesis.typepad.com/synthesis/2006/05/synthetic_biolo_1.html (accessed 1 Dec 2008).
Cello, J., A.V. Paul and E. Wimmer, (2002), 'Chemical Synthesis of Poliovirus cDNA: Generation of Infectious Virus in the Absence of Natural Template', *Science*, 297: 1016–1018.
Chan, L.Y., S. Kosuri and D. Endy, (2005), 'Refactoring bacteriophage T7', *Mol Sys Biol*, DOI: 10.1038/msb4100025.
Chin, J., (2008), 'Life and perpetuation of life of a synthetic bacterium', EMBL/EMBO Science and Society Conference, Heidelberg, Germany, 7th-8th November 2008.
Conley, J.M. and R. Makowski, (2004), 'Rethinking the product of nature doctrine as a barrier to biotechnology patents in the United States – and perhaps Europe as well', *Information & Communications Technology Law*, 13(1): 3–40
Deamer, D., (2005), 'A giant step towards artificial life?' *Trends in Biotechnology*, 23: 336–338.
Dueber, J.E., B.J. Yeh, K. Chak and W.A. Lim, (2003), 'Reprogramming control of an allosteric signaling switch through modular recombination', *Science*, 301: 1904–1908.
Dupré, J., (forthcoming 2010), 'Is it not possible to reduce biological explanations to explanations in chemistry and/or physics', in Ayala, F. and R. Arp (eds), *Contemporary Debates in Philosophy of Biology*, Hoboken, NJ: Wiley.
Dutfield, G., (2008) 'Who invents life – intelligent designers, blind watchmakers, or genetic engineers?' SIBLE Seminar Series, University of Sheffield, 5 March 2008.
Elowitz, M.B. and S. Leibler, (2000), 'A Synthetic Oscillatory Network of Transcriptional Regulators'. *Nature*, Jan 20; 403(6767): 335–338.
Endy, D., (2004), 'Parts, devices & systems: engineering biology at MIT', *Synthetic Biology* 1.0, MIT, 10th–12th June 2004.
Endy, D., (2005), 'Foundations for Engineering Biology', *Nature*, 438, 24 November: 449–453.
Ferber, D., (2004), 'Microbes made to order', *Science*, 303, 9 January: 158–161.
Franklin, S., (2000), 'Life Itself: Global Nature and the Genetic Imaginary', in Franklin, S., C. Lury and J. Stacey (eds), *Global Nature, Global Culture*, London: Sage: 188–227.
Franklin, S., (2003), 'Kinship, genes, and cloning: Life after Dolly', in Goodman, A., D. Heath and S. Lindee (eds), *Genetic Nature/Culture: Anthropology and Science Beyond the Two Culture Divide*, Berkeley: University of California Press: 95–110.
Franklin, S., (2006), 'The Cyborg Embryo: Our Path to Transbiology', *Theory, Culture and Society*, 23: 167–187.
Gallivan, J.P., (2007), 'Toward reprogramming bacteria with small molecules and RNA', *Curr Opin Chem Biol*. 2007 December, 11(6): 612–619.
Glass, J.I., N. Assad-Garcia, N. Alperovich, S. Yooseph, M.R. Lewis, M. Ma-ruf, C.A. Hutchison III, H.O. Smith and J.C. Venter, (2006) 'Essential Genes of a Minimal Bacterium', *Proc Nat Acad Sci*, 103(2): 425–430.
Griesemer, J.R. and Gerson, E.M., 'Essay review. Of mice and men and low unit cost', (review of Making mice: Standardizing animals for American biomedical research, 1900–1955, Karen A.

Rader; Princeton University Press, Princeton, NJ, 2004), *Stud. Hist. Phil. Biol. & Biomed. Sci*, 37(2006): 363–372.

Hacking, I., (1992), 'The Self-Vindication of the Laboratory Sciences', in Pickering, A. (ed.), *Science as Practice and Culture*, Chicago and London: Chicago University Press: 29–64.

Hartwell, L.H., J.J. Hopfield, S. Leibler and A.W. Murray, (1999), 'From molecular to modular cell biology', *Nature*, 402: C47–C52.

Heinemann, M. and S. Panke, (2006), 'Synthetic biology – putting engineering into biology', *Bioinformatics*, 22(22): 2790–2799.

Highfield, R., (2008), '"Watermarks" written in first artificial genome', *Telegraph*, 01 Feb 2008.

Isaacs, F.J. and J.J. Collins, (2005), 'Plug and play with RNA', *Nature Biotechnology*, 23(3): 306–307.

Jasanoff, S., (2006), *States of Knowledge: The Co-Production of Science and the Social Order*, New York: Routledge.

Keasling, J., (2009), 'Keynote lecture' Launch Event for the Centre of Synthetic Biology at Imperial College, 12th May 2009.

Keasling, J.D. and H. Chou, (2008), 'Metabolic engineering delivers next-generation biofuels', *Nature Biotechnology*, 26: 298–299.

Keller, E.F., (2008a), 'Lecture: nature and the natural', *BioSocieties*, 3: 117–124.

Keller, E.F., (2008b), 'Systems biology: new paradigm or just fashion', EMBL/EMBO Science and Society Conference, Heidelberg, Germany, 7th-8th November 2008.

Kevles, D.J., (2002), *A History of Patenting Life in the United States with Comparative Attention to Europe and Canada* A report to the European Group on Ethics in Science and New Technologies, 12th Jan 2002 Luxembourg: Office for Official Publications of the European Communities,

Kitano, H., (2002), 'Computational systems biology', *Nature*, 420, 14th November 2002: 206–210.

Kitney, R., (2008), 'Overview of Design Principles', Synthetic Biology 4.0, Hong Kong, 10th–12th October 2008.

Kohler, R.E., (1999), 'Moral economy, material culture, and community in Drosophila genetics' in Biagioli, M. (ed.), *The Science Studies Reader*, New York: Routledge.

Lancet, D., (2008), 'Diversity: a driving force for life's inception and synthesis', EMBL/EMBO Science and Society Conference, Heidelberg, Germany, 7th–8th November 2008.

Lazebnik, Y., (2002), 'Can a biologist fix a radio?–Or, what I learned while studying apoptosis', *Cancer Cell*, Vol. 2: 179–182.

Lentzos, F., G. Bennet, J. Boeke, D. Endy and P. Rabinow, (2008), 'Roundtable on synthetic biology: visions and challenges in redesigning life', *BioSocieties*, 3: 311–323.

Luisi, P.L., F. Ferri and S. Pasquale, (2006), 'Approaches to Semi-synthetic Minimal Cells: A Review', *Naturwissenschaften*, 93(1): 1–13.

Loettgers, A., (2007), 'Model Organisms and Mathematical and Synthetic Models to Explore Gene Regulation Mechanisms', *Biological Theory* Spring 2007, Vol. 2, No. 2: 134–142.

Martin, V.J.J., D.J. Pitera, S.T. Withers, J.D. Newman and J.D. Keasling, (2003), 'Engineering a mevalonate pathway in Escherichia coli for production of terpenoids', *Nature Biotechnology*, 21: 796–802.

Martins dos Santos, V., (2008), 'Powering Cell Factories', *Synthetic Biology*, 4.0, Hong Kong, 10th–12th October.

Marcus, G.E., (1995), 'Ethnography in/of the World System: The Emergence of Multi-Sited Ethnography', *Annual Review of Anthropology*, Vol. 24: 95–117.

Morange, M., (2009), 'A new revolution? The place of systems biology and synthetic biology in the history of biology', *EMBO reports*, Vol. 10: S50–S53.

NEST, (2005), *Synthetic Biology: Applying Engineering to Biology*, Report of a NEST High-Level Expert Group Luxembourg: Office for Official Publications of the European Communities online at ftp://ftp.cordis.europa.eu/pub/nest/docs/syntheticbiology_b5_eur21796_en.pdf (accessed 14 Sep 2008).

Oliver, S., (2008), 'Synthetic and Systems Biology: Simplicity and Simplification', Discussion Meeting on Synthetic Biology, Royal Society, London, 2nd–3rd June 2008.

O'Malley, M.A., A. Powell, J.F. Davies and J. Calvert, (2008), 'Knowledge-making distinctions in synthetic biology', *BioEssays*, 30(1): 57–65.

Palsson, B., (2000), 'The challenges of in silico biology', *Nature Biotechnology*, Vol 18 November, 2000: 1147–1150.

Pollack, A., (2001), 'Scientists Are Starting to Add Letters to Life's Alphabet', *New York Times*, 24 July.

Pottage, A., (1998), 'The inscription of life in law: genes, patents, and bio-politics', *Modern Law Review*, 61: 740–765.

Pottage, A., (2007), 'The socio-legal implication of the new biotechnologies', *Annual Review of Law and Social Science*, 3: 321–344.

Pottage, A., (2009), 'La mécanisation propriétaire du vivant', La co-définition du vivant et des droits de propriété, Paris, 9 April 2009.

Pottage, A. and B. Sherman, (2007), 'Organisms and manufactures: on the history of plant inventions', *Melbourne University Law Review*, 31(2): 539–568.

Rabinow, P., (1999 [1992]), 'Artificiality and Enlightenment: From Sociobiology to Biosociality', in Biagioli, M. (ed.), *The Science Studies Reader*, New York: Routledge: 407–416.

Rheinberger, H-J., (2000), 'Beyond Nature and Culture: Modes of Reasoning in the Age of Molecular Biology and Medicine', in Lock, M., A. Young and A. Cambrosio (eds), *Living and Working with the New Medical Technologies: Intersection of Inquiry*, Cambridge: Cambridge University Press: 19–30.

Royal Academy of Engineering, (2009), *Synthetic Biology: Scope, Applications and Implications*, Royal Academy of Engineering: London, May 2009, available online at http://www.raeng.org.uk/news/publications/list/reports/Synthetic_biology.pdf (accessed 5 July 2009).

Sauro, H.M., (2008), 'Modularity Defined', *Molecular Systems Biology*, 4: 166.

Silver, P., (2008), 'Designing Biological Systems', Discussion Meeting on Synthetic Biology, London: Royal Society, 2nd–3rd June 2009.

Silver, P., (2009), 'Design Principles in Biological Systems', Launch Event for the Centre of Synthetic Biology at Imperial College, 12th May 2009.

Tucker, J.B. and R.A. Zilinskas, (2006), 'The promise and perils of synthetic biology', *New Atlantis*, 12: 25–45.

Villalobos, A., J.E. Ness, C. Gustafsson, J. Minshull and S. Govindarajan, (2006), 'Gene Designer: a synthetic biology tool for constructing artificial DNA segments', *BMC Bioinformatics*, 7: 285 doi:10.1186/1471-2105-7-285.

De Vriend, H., (2006), *Constructing Life. Early social reflections on the emerging field of synthetic biology*, The Hague: Rathenau Institute; Working Document 97, available online at http://www.rathenauinstituut.com//showpage.asp?steID=2&item=2644 (accessed 25 Feb 2008).

Weiss, R., (2008), 'The iGEM Undergraduate Competition', *Synthetic Biology* 40, Hong Kong, 10th–12th October 2008.

Wimsatt, W., (2007), *Re-engineering philosophy for limited beings: piecewise approximations to reality*, Cambridge, MA: Harvard University Press.

Wilmut, I., K. Campbell and C. Tudge, (2000), *The Second Creation: The Age of Biological Control by the Scientists Who Cloned Dolly*, London: Headline.

Wynne, B., (2005), 'Reflexing complexity: post-genomic knowledge and reductionist returns in public science', *Theory, Culture and Society*, 22(5): 67–94.

Interspecies entities and the politics of nature

Sarah Parry

> Ways of living and dying matter: Which historically situated practices of multispecies living and dying should flourish? There is no outside from which to answer this mandatory question; we must give the best answers we come to know how to articulate, and take action, without the god trick of self-certainty (Haraway, 2008: 88).

Introduction

The UK *Human Fertilization and Embryology Authority* (HFEA) recently received two research applications to create animal-human embryos. These license requests hailed from two separate research teams at the University of Newcastle and King's College London. Both applications proposed to create embryos using non-human enucleated oocytes (a non-human egg with its nucleus removed) with the nucleus from a human cell using a technique known as somatic cell nuclear transfer.[1] If this research is successful they hope to derive stem cell lines from the resulting animal-human embryo, which, in this chapter, I will call an interspecies entity.

As one might expect, these license applications generated a number of political and public debates in the UK concerning the social, ethical, political and regulatory implications associated with the creation of these interspecies entities (see Brown, 2009). However, the debate extended beyond the case of animal-human embryos to include other kinds of interspecies entities where ideas of 'humanness' and 'animalness' are also challenged. Other examples included so-called 'true hybrids' created from the sperm of one species and the egg of another, and 'chimeras' where cells from one species are transferred into the developing embryo of another. These are but two of the myriad forms of interspecies entities produced through xeno-technologies (see Brown and Michael, 2001; Fox, 2005) where matter from one species is purposefully introduced into another. These new or potential technoscientific developments at one and the same time offer new possibilities for creating life and challenge established ideas of species and the relations between them. These concerns were taken up in the public policy-related debates which focussed on establishing meanings and definitions of these different entities in order to then establish the UK's national regulatory framework for their creation and use in scientific and medical research (see Brown, 2009).

Given the ferocity of these debates, it is not surprising that before issuing or declining these license requests the HFEA initiated its largest public consultation process yet in the early summer of 2007. During this period, I was leading an ESRC-funded project, *The Social Dynamics of Public Engagement in Stem Cell Research*.[2] Taking our cue from these new research developments and public debates in stem cell research, we organized a small public debate of our own on the topic. We brought together 17 participants from diverse backgrounds, including: stem cell scientists, members of patient groups, nurses, interested members of the public, anti-abortion campaigners, social scientists and bioethicists, to encourage discussion of the complex issues involved and to make their own contribution to the HFEA consultation. For the sociologist, these debates offer important insights into the multiple discourses about animals, humans and the relationship between the two that are circulating today. But more than this, and in keeping with the spirit of the excerpt from Haraway used to open this chapter, from here we can consider the implications of these discourses – both material and discursive – for different human and non-human animals. It is these two concerns that provide the focus for this chapter: how are ideas of animal and human nature figured in the case of interspecies entities and what are the material and discursive implications of these inscriptions?

Animal-human relations in context

The classification of animals and humans and the relationship between them has generated much research. However this chapter will not attempt to review a literature that spans disciplines as varied as sociology, philosophy, politics, geography and gender studies. Nevertheless, it is important to note some key ideas from this literature in order to provide conceptual lineages, context and of course intellectual acknowledgement for what follows.

The first is that the classification of the human and the non-human is neither universal, fixed in time, or singular. Instead, classificatory systems are culturally and historically specific, and open to negotiation. John Dupré's (1993, 2002) concept of 'promiscuous realism' is particularly instructive here in highlighting that taxonomies are not underpinned by fixed essences that can be identified and used as criteria for membership of, say, species. Rather, what determines the conferral of membership of an individual entity to a particular species or other taxonomical class depends 'on both the purposes of the classification and the peculiarities of the organisms in question, whether those purposes belong to what is traditionally considered part of science or part of ordinary life' (Dupré, 1993: 57). That is, any given entity may be (and indeed, is likely to be) classified in more than one way and the principles of classification propounded by biological scientists should not be accorded greater legitimacy than those in other domains (eg lay publics). We must acknowledge that humans select and privilege some characteristics over others when classifying entities and that classificatory systems, as a social accomplishment, are underdetermined by nature (see Barnes, 1983; see also Marks, this volume). Hence, an awareness of the inherently political nature of classification is brought to the fore.[3] Dupré also reminds us to attend to the purposes of classification and so, in the context of interspecies entities, we might paraphrase Bowker and Star's (1999) statement 'to classify is human', and postulate 'to classify is political'.

Second, we need to take into account the relationship between different classes of entities, particularly human and non-human animals. It has long since been acknowledged that the 'human' is too often defined in terms of what it is not – an Other – where the 'non-human' refers to the 'animal' (see Haraway, 2008). Indeed, much of the humanist agenda (see Twine, this volume) has involved separating, essentializing and bounding categories of human and non-human animals, and is marked by discourses of human exceptionalism that have all too often served to legitimate certain forms of oppression and exploitation of non-human animals. In an attempt to move beyond dualist understandings of animals and humans, and the associated humanist assumptions, a number of writers have advocated an approach that captures the interdependence and interplay of human and non-human animals.

Agamben's eloquent exposition of the relationship between the animal and human in *The Open* (2004) has provided valuable insights into the border between the human and non-human – what he calls the 'zone of indifference'. These zones are populated by humanized animals (such as the ape-man) and animalized humans (such as slaves), where the symbolic distinction between human and non-human animals is not clear cut but is more accurately described as graduations of animal or human. It is this relational constitution that, according to Agamben, has enabled and legitimated oppressive and often violent acts against those in this zone. Notably, it is the animalization of humans that has, at different points in history, legitimated practices such as slavery. In this sense, and importantly for this chapter, Agamben argues that the production of the human and non-human binary or distinction is central to politics itself.

While calling our attention to the relational aspects of animal-human classification, Agamben's attention is firmly fixed on the implications for humans (see Davies, this volume). Haraway (2008), on the other hand, in her relational understanding of human and non-human animals redresses this balance in her book *When Species Meet*. Like Agamben, Haraway's work is politically engaged and is itself a political project. She similarly seeks to disrupt pernicious forms of humanism by undoing:

> . . . the discursive tie between the colonized, the enslaved, the noncitizen, and the animal – all reduced to type, all Others to rational man, and all essential to his bright constitution (2008: 18).

But a key difference here is Haraway's attempt at a symmetrical appreciation of the implications of our material and discursive practices for all life, whether defined as human or otherwise. In focussing on spaces where humans and animals encounter one another, such as the laboratory, Haraway does not treat species as stable categories but instead is concerned with processes of becoming at both the epistemological and ontological levels.

But these discussions about how human and non-human animals are relationally constituted are culturally situated in a context of profound contradiction in how we both understand and treat animals in contemporary society. On the one hand, we have a prevalent discourse of 'rights' with associated legislation and institutions to protect those who fall under their remit, including responsibilities for pet owners and animals held in laboratories. Linked to this are the close relations humans form with animals as pets. On the other hand, this co-exists with ongoing human practices where animals continue to be used for food and in scientific and medical research. Benton and Redfearn have described this contradiction as the gulf between what they call the 'growing conviviality' and the 'intensifying reification' of animals (1996: 49). In particular, they are interested in the cultural specificity of animal-human relations (eg the eating of snails but not horse meat in France) and also the cultural mechanisms that prevent 'our compassion for the members of one animal species . . . from seeping out to affect feelings for another' (Benton and Redfearn, 1996: 49).

Benton and Redfearn are concerned with unveiling the various techniques of concealment with regard to the exploitation of animals, and align themselves with the concept of 'animal rights'. In contrast, Haraway neither adopts this position nor advocates a principled position against the use of non-human animals in scientific and medical research. She does, however, critique the way in which the killing of animals (and humans) is currently practised as a form of industrialized, commodified killing that fails to appreciate that 'we are in a knot of species coshaping one another' (2008: 42). Instead, Haraway argues, we must develop more responsible understandings of the 'living and dying, and nurturing and killing' (2008: 42) of human and non-human animals that starts with an appreciation of their always asymmetrical relations. And, normatively speaking, this refers to how humans treat not only animals but other humans too.

Taking these ideas forward to make sense of the multiple discourses circulating around interspecies entities poses an intriguing challenge. This is to ascertain how animals, humans and their relations are actually figured in these debates. While the data that is drawn on must also be considered in terms of its limitations (ie it is a limited sample, drawn from a particular moment in time and space) it remains possible to discern a sense of the range of discourses being articulated. From here, we must then consider the material and political implications of such discourses because, as the above discussion has indicated, classification is not an innocent process but is inherently political.

Interspecies entities in context

This chapter is located in both an ongoing set of redefinitions of animal-human relations and their material reconstruction. The proposed creation of animal-human embryos involves *creating* entities that do not exist in nature in a material sense. Only through technoscientific intervention and laboratory practices can these interspecies entities materially come into being. However, as I will first explain, their creation cannot be understood in isolation but instead must be located in the more complicated setting of stem cell research where human and non-human animals are symbolically and materially relational.

Before the proposals to create animal-human embryos, scientists who wanted to generate new human embryonic stem cell lines were reliant upon the use of tissue obtained from women's bodies – their eggs and/or embryos. Since changes to the 1990 *Human Fertilization and Embryology Act* (HFE Act) were implemented in 2001, stem cell scientists have been permitted to recruit women undergoing IVF to donate their 'failed-to-fertilize' eggs and/or 'spare' embryos for research purposes. But as the embryonic stem cell research field has grown over the last decade, so too has the requirement for embryos and, particularly, eggs by researchers. As scientists try to further their understanding and development of somatic cell nuclear transfer, eggs have become highly valued entities. However, failed-to-fertilize eggs have not proved to be a very successful material either for creating cloned embryos or deriving stem cell lines. A combination of these poor success rates and increasing demand fuelled by this growing field of research, has resulted in recent regulatory changes to expand the recruitment of human egg donors for stem cell research (cf. Roberts and Throsby, 2007). Further changes to the HFE Act in 2007 now permit the recruitment of egg donors for stem cell research through both (so called) 'egg sharing' arrangements (where women undergoing IVF donate half of their eggs in return for reduced treatment fees) and donation from compensated non-patient women donors (ie from women not undergoing IVF who, in return for some eggs, receive a small amount of financial compensation) (Roberts and Throsby, 2007). The aim here was to obtain greater numbers of 'healthy' fresh eggs in order to improve somatic cell nuclear transfer and try to generate stem cell lines. Taken as a whole, it is clear that women's bodies constitute a material resource for stem cell research that also has a regulatory mandate in UK (the HFEA).

The use of human eggs and embryos in stem cell research has been continually subjected to public scrutiny and ethical debate; particularly, in the public domain, by 'pro-life' campaigners, who argue that somatic cell nuclear transfer involves the exploitation and destruction of human life for the purposes of conducting experimental research. Stem cell scientists who use human eggs and/or embryos in their research, therefore, have had to engage in ongoing and sometimes highly visible public debates to continually (re)establish the technical and ethical legitimacy of this research (cf. Parry, 2009). It is this context of debates regarding the use of *human* eggs and embryos in stem cell research in which the use of *non-human* eggs was proposed, and, therefore, provided the starting point for the HFEA's public consultation.

The document supplied by the HFEA as background for the public consultation, *Hybrids and Chimeras* (2007), locates the use of non-human eggs as an ethically preferable alternative to using women's. Recruitment of women egg donors (through egg sharing and compensated non-patient donors), it states, has been difficult 'largely because donation is a physically demanding process which can, in rare cases, harm the donor's health'. Similarly, Stephen Minger, one of the scientists from Kings College London who applied for a licence to create interspecies entities, has been a vocal supporter of creating interspecies embryos as an alternative to using women's eggs. Minger publicly opposed the recruitment of women egg donors through both egg sharing and compensated non-patient donation, on the grounds that somatic cell nuclear transfer remains unproven, the number of eggs required to produce a single stem cell line is expected to run into the thousands, and there are significant health risks for the women with no medical benefit (see written evidence from Stephen Minger, Memorandum 49, House of Commons Science and Technology Committee, 2007). The use of non-human eggs in somatic cell nuclear transfer is not presented as so ethically contentious and thus some, such as Minger, sees the use of non-human eggs as a legitimate response to the reputed scarcity of human eggs being donated for stem cell research.

The public discussion of interspecies entities which was run as part of the project, *The Social Dynamics of Public Engagement in Stem Cell Research*, also contextualized these new developments in debates about existing sources of *human* eggs. In this respect, we drew upon discussions taking place in public and policy domains as starting points for our debate, and asked participants to consider and share their views on three substantive issues: i) existing *human* sources of eggs and embryos for stem cell research, ii) the creation of interspecies *embryos*, and iii) the creation of *other* kinds of interspecies entities. These were discussed in three focus groups. Each group contained a stem cell scientist and up to six participants.[4] All of the groups were tape-recorded and the discussions transcribed and analysed.

In the following section I describe four discourses of human and non-human nature that were drawn upon during the focus groups. As I will go on to discuss in the concluding section, these discourses coexist and yet have differing material (ontological) and epistemological implications for human and non-human animals.

Debating interspecies entities

> [biotechnologies are] . . . generating messmates at [the] table who do not know how to eat well and in my judgement often should not be guests together at all. Which companion species will, and should, live and die, and how, is at stake (Haraway, 2008: 17–18).

The creation of interspecies entities for stem cell research involves an unprecedented yet unequal involvement between human and non-human animals. In the very creation of these novel entities, decisions are being made about who lives and who dies. However there is little consensus when it comes to views about the living and dying involved either within the community of stem cell scientists or society more broadly. Similarly, there is little consensus regarding what these interspecies entities mean. Hence, my concern in the following section is to unpack the underpinning views about humans, non-human animals and their relations: how, according to what criteria, and to what extent are interspecies entities differently understood? Further, what can this tell us about how we understand human and non-human animals?

Human and non-human animal distinctions are inviolable

The first discourse to emerge in the group discussions concerns the idea that human and non-human animals are understood as distinct beings with fixed essences. Moreover, these essences and the distinction between human and non-human animals, according to this view, should not be undermined by innovations in science, technology and medicine. Both the pro-life campaigners and the research nurses who participated in the groups put this view forward in the public debate.

As already discussed, using non-human eggs instead of human eggs was framed by some as an ethically preferable route to developing cloned embryos using somatic cell nuclear transfer. One assumption here (and also echoed by some stem cell scientists at our public debate – I will return to this later), is that groups, such as *Life*, who have been vocal in their opposition to the creation of all human embryos for research purposes, will find the creation of interspecies entities acceptable. However, the pro-life campaigners who participated in our debate argued against any research that involves introducing biological material from one species into another (whether animal-to-human or vice versa). When asked about the move towards creating embryos using non-human eggs one pro-life campaigner argued it is 'more' troubling than using women's eggs because 'anything which has any human DNA in it' is recognized as human. Therefore, she argued, research involving the creation of interspecies entities is evidence of further commodification and disrespect towards early *human* life. In the context of human–non-human animal relations, the elaboration of this position is particularly interesting.

Creating a life using both human and non-human genetic material unsettles what is considered distinctive about humans – cognitive capacities and the 'divinity within . . . [a] . . . human being that is not there in animals' (pro-life campaigner A), ie traits which are not deemed to be present in non-human animals – and begins to blur the boundary between the human and non-human. This view was argued to apply not only to the creation of interspecies embryos but also to other practices that transfer genetic material from humans to non-human animals and vice versa:

> I would probably say we would take issue with injecting human DNA into animals just because we would just see that as morally and ethically wrong in terms of [the] disrespect to what we would consider particularly special and unique to human beings. [This] is not what we share with animals (Pro-life campaigner A).

Underpinning these accounts about the distinctions between humans and non-human animals is the notion that we can locate essences that are possessed by humans in order then to identify humans as human and not animal. The animal, here, is identified through what it lacks rather than what it possesses. Here, the view that inserting human genetic material into an animal reveals further disrespect for human life simultaneously sustains the position that human life carries more value than animal life.

Research nurses also opposed inserting human genetic material into non-human animals, or vice versa. Like the pro-life campaigners, they too held a principled objection to the creation of all interspecies entities:

> Nurse A: But that's the bit that makes me uncomfortable.
>
> Moderator: Does it?
>
> Nurse A: Yes, it's that crossing. The humans-to-humans that's fine, muck around, sort it out. You know, animals-to-animals, that's fine too. But if you cross over, mule, I mean, donkeys and horses make mules, there's something not quite . . .

The research nurses' views must be placed in context. In their professional lives, one of their roles is to obtain consent from women undergoing surgery (such as abortions or 'sterilization') or IVF to access their fetuses or eggs and/or embryos respectively, which will then be used in research (stem cell and other). One of the most notable ways in which this context impacted upon their explanations of their objections to creating interspecies entities was through the emphasis they placed on the importance of consent. Women, they argued, can give consent to donate their eggs whereas non-human animals cannot:

> you may be doing invasive stuff to them [women], but you know, it's still their decision . . . And the first time I was involved in it and went into theatre and saw it being done, and I thought, oh my God, what am I doing involved in this? You know, this is quite an invasive procedure and things. It's like anything, you kind of get used to it. And the patients go through quite a heavy consenting procedure. We go through everything in detail with them and explain all the risks and no benefits to them apart from they've got an altruistic reason to donate (Nurse B).

This was a view shared by other participants, including somebody with experience as a bench-scientist who chose to work with human tissue for animal welfare reasons. But when Nurse B was pushed by another participant as to whether it is the animal welfare issues that fundamentally concerns her, the answer was partly yes but perhaps more 'just the animal-human cross' (Nurse B). Implied in this account is something similar to the pro-life campaigners views – humans possess cognitive capacities and essences that are not shared with non-humans. The description of women's ability to give consent provides a narrative through which they could highlight how interspecies entities transgress or even undermine something of the human essence.

A final dimension to the inviolability of human and non-human animal nature concerns the type of violation – some interspecies entities undermine the human essence whereas others do not. Here, non-human animals considered closer to the human, such as non-human primates, are argued to be unacceptable sources of biological material to be mixed with humans by a number of participants, particularly the stem cell scientists. Non-human animals considered more different, such as rabbits and cows, however, do not constitute the same cultural risk to our sense of humanness:

> I don't have a problem with that [using rabbit eggs], as I say it's only when they go higher up the chain, you know, this thing about consciousness and sentience and all that. When you go higher up the chain that's when I start having a problem. I have no problem whatsoever with research with rabbits, rats or whatever, you know, that's not a problem. It's when you get to this closeness between us and other species, that's where I have a problem (Clinical ethics researcher).

Here, what is considered a human essence is also considered to be shared by non-human primates: 'that's quite frightening because sentience and consciousness are almost identical' (clinical ethics researcher). As in Agamben's analysis, it is in these spaces of close proximity where much of the cultural, social and political work is in operation (and has been so historically), that some of the greatest felt threats to the human are located.

Human and non-human animal nature as materially dynamic

A second discourse establishes human and non-human animal nature as materially dynamic, emphasizing how identity is open to shifting between the two. What constitutes a human or non-human animal is not fixed at the material level and can change over time, particularly through technoscientific and medical intervention. In an interspecies context – a context where both human and non-human biological material co-exists – the material matter (the cells) does not remain fixed as *inter*species, but shifts (or switches) over time to become human or otherwise. This material shift is coupled to an epistemological shift, particularly with regard to eschewing species indeterminacy or a third *inter* category (see next section).

One stem cell scientist provided a striking example of this when he described how the stem cells generated from interspecies embryos created using a rabbit's egg with a nucleus from a human stem cell line become increasingly human over time:

> . . . So, the cytoplasm [the material from the rabbit] gets diluted in time because the nucleus produces, has all the genetic material, and as that cell divides, every division, it becomes that much less a rabbit cytoplasm and that much more a human cell (Stem cell scientist B).

This account of the increasingly humanized cell line, where the human 'takes over' the animal, has also featured in the policy-related debates in 2007 (see

Brown, 2009). It is this dynamic understanding of what it is to be 'animal' and 'human' that helps establish the view that interspecies embryos are 'basically human' (Stem cell scientist B). As Brown (2009) has already discussed, it is the primacy given to the nucleus that underpins and therefore, often determines the status of an entity as human or non-human animal. Significantly, it is this 'speciesist twist' (Brown, 2009: 159) in the interspecies entities debate aligning nucleus, human, permanent against mitochondria, animal, impermanent that is then used to argue for the ethical preference for using *non-human* eggs to create *human* embryos.[5]

Although interspecies embryos are considered to begin as animal but then become human because their materiality is transformed (the animal mitochondria is taken over by the human nucleus), this is not considered to be the case for all interspecies entities. Participants reflected on whether the introduction of all types of cells from one species to another affects a shift in species identity. For example, the transplantation of non-human organs into humans, it was argued, does not affect species identity: a human with a pig's heart or kidney indisputably remains a human. Similarly, putting human heart cells into a non-human to test for therapeutic efficacy (ie do they function in the animal body as they ought?) was equally unproblematic in terms of shifting the identity of that non-human animal. In contrast, inserting functioning human brain cells into a non-human animal was argued to threaten a shift in identity where the (hypothetical) non-human animal becomes increasingly humanized, or indeed human:

> Yes, is that mouse then going to kind of become human because its brain cells, what makes us human I suppose, is now, could be in a mouse (Member of regulatory body).

Therefore, in this discourse, our knowledge and representation of species identity is coupled to and indeed driven by what is understood to be occurring at the material level. Some kinds of animal cells can affect a transformation while others cannot.

Human and non-human animal natures

The above discussion begins to pre-empt a third discourse that emphasizes a more plural epistemology of animal and human natures. Here, the material and epistemological entanglement of human and non-human animal natures creates a space for the multiple possibilities for meaning regarding what constitutes a human or non-human animal. Further, ideas of material purity are less important in this discourse. Yet, this multiplicity is not limitless, as I discuss below.

Like the example of animal-to-human organ transplantation discussed above, some participants argued that although an interspecies embryo may contain a small percentage of non-human genetic material, the entity is nonetheless always understood as a human:

Stem cell scientist C: If they took a rabbit embryo and put a human nucleus in and eventually get a cell line, that will be essentially a human cell line.

Moderator: It won't still have the 1 per cent mitochondria?

Stem cell scientist C: It will still have the rabbit mitochondria.

Moderator: Yeah, so that's a definition.

Unlike the previous section, where the human nucleus was constructed as taking over the non-human mitochondria, in this account the non-human content co-exists with the human and yet does not ever overwhelm the human identity. Instead, the material co-existence of human and non-human genetic material raised questions for the group about the meaning of numbers in relation to human and non-human genetic content – a question the focus group considered difficult to resolve even through further scientific research:

> Science might say its 99 per cent human, how do you test that? How do you actually know that this is human? Surely the thing about being human is about consciousness and sentience and all the rest of it. You're never going to test that because it's never going to go full term. So, the whole definition around is it human is it not, to me, is problematic because you're never actually going to know (Clinical ethics researcher).

Hence, while genes might be the 'theoretico-empirical sign of the Big Gap popular at the moment' (Haraway, 2008: 79) for distinguishing humans from non-human animals, our discussants acknowledged that this has limitations when it comes to definition-making (cf. Brown, 2009; Holmberg, 2005; Marks, 2002). Instead, we see the interplay between different kinds of essences (eg cognitive or genetic) and how they might be used to determine the classification of an entity that perhaps doesn't include recourse to notions of 'purity' or even certainty. In this sense, there was a pragmatic ethics in play in these discussions where some participants were willing to define entities as human or otherwise in a way that resonates with a situated understanding to knowledge claims (cf. Haraway, 1991). Further, notions of species purity at the material level are less important in this discourse.

Others were more explicitly comfortable with a third, or inbetween, category. One stem cell scientist in particular, eschewed the human–non-human animal distinction altogether. When asked about creating embryos using a non-human animal egg with a human nucleus she argued that the resulting entity is not a *human* embryo '. . . because they have an animal component to them, so no I wouldn't [regard them as human]'. The moderator then uses a thought experiment to explore the point further by asking the groups' view of creating an embryo with a human egg and a non-human animal nucleus:

Moderator: . . . does that make it any different as to whether that's a human or an animal?

Stem cell scientist A: Yes to me it would. It would still be neither. Therefore even in that form, I can't, well I can't see the point in anybody ever conceiving that thought.

What emerges from this discussion is that the stem cell scientist sees no need to define interspecies embryos as either human or non-human animals. The discussion that followed in the group showed that others also wrestled with how to define these entities and a great deal of uncertainty regarding their human or non-human status. However, for stem cell scientist A, one key point in defining these entities as other-than-human is to get 'round the ethics' of using women's eggs or spare IVF embryos in stem cell research. Indeed, she said that, as a woman, she wouldn't donate her eggs to stem cell research 'Because it's such an invasive technique and I worked in the NHS for a long time and I know what's [involved]' (Stem cell scientist A).

The obduracy and exuberance of animal and human nature

The final discourse operating in our public debate positioned nature as obdurate or sometimes exuberant. Rather than being a compliant collaborator in stem cell research, the materiality of nature is seen to undermine and resist the efforts of scientists in various ways. In this sense, nature resists being moulded into the visions of human actors resulting sometimes in no change (obduracy) and other times unexpected change (exuberance).

One common example has been the creation of so-called true hybrids. These are entities created using the egg from one species with the sperm of another such as donkey-horse (mule or hinny) hybrids. In discussion, these are raised as examples that while life forms *can* be created from two species they are also obdurate and exuberant:

> It's interesting that if you cross a horse with a donkey it's infertile. It's like it's saying, 'don't do that' (Nurse A).

Along similar, but more technical, lines others described how two species that are too different, too distant, will resist attempts to create life:

> Stem cell scientist C: A lot of it won't work – two species of two different, you are going to get a few divisions.
>
> Moderator: The whole mechanism of the cell.
>
> Stem cell scientist C: It will just be a mess, especially if they have unequal numbers of chromosomes. The chromosomes have to line up to separate.
>
> Biotechnology entrepreneur: Yeah.
>
> Stem cell scientist C: Maybe a non-[human] primate and the human.
>
> Moderator: Or a cloned species, yeah?
>
> Stem cell scientist C: But a mouse sperm and a human egg, that's probably going to die after two divisions or something.

This discussion was also echoed in other groups.

The perceived distance between two species plays out in a further twist in this discourse of obdurate and exuberant nature. It is not only the obduracy of the chromosomal materiality of human and non-human animals that eludes scientists' efforts but also an observable materiality to the non-scientifically trained eye: the physical form. Using the example of mice and elephants, one clinical ethicist argued that if the physical form is observably too different then this is both materially and epistemologically troubling. Here, the spectre of Frankenstein's monster entered our public debate to illustrate this is a step too far. Thus, the clear, observable material difference in terms of size between these two animals is invoked as imposing a brake on research that might mix their biological material.

A final example of the exuberance of nature arises in terms of the risk of transmitting viruses from one species to another through this research. One group in particular focussed some of their discussion on the limited therapeutic uses of cell lines created from interspecies entities because of these risks. Likened to the risks of 'Mad Cows Disease' or variant-Creutzfeldt-Jakob Disease, transferring cells from interspecies entities to humans as part of a therapy was deemed unlikely. As nurse B goes on to say: 'We all move heaven and earth to remove animal products from stem cell production in order to be able to use them for cell therapy'. Here, these risks represent the exuberance of nature, which is seen to resist the agenda of some scientists – nature is non-compliant.

Concluding discussion

The focus for this chapter has been to examine how ideas of animal and human nature are figured in the case of interspecies entities and to ask what are the material and discursive implications of these inscriptions. In analysing the accounts produced in the focus groups that took place in this public event, it is evident that as participants actively wrestle with the issues and implications of interspecies entities, they simultaneously assemble new ways of thinking and speaking about such things. Here, these technoscientific possibilities shape and are shaped by people's sensibilities. The four discourses of human and non-human animal nature presented above imply different consequences for humans and others. And while I have presented them as somewhat distinct discourses, these must not be treated as either/or accounts but instead coexist, with participants moving between them as they attempt to make sense of interspecies entities. In the following discussion, I will tease out some implications of these discourses while also highlighting areas for further research.

In the first discourse – *human and non-human animal distinctions are inviolable* – we can observe a partial rehearsal of a humanist agenda. Here, animals (whether human or otherwise) have fixed essences (and therefore fixed boundaries) that should not be undermined or transformed by technoscientific research practices. However, whereas other critics of humanism have pointed to its role in legitimating research using non-human animals, in the interspecies context we see an interesting twist. As illustrated in the pro-life accounts, the definition of such entities as *human* designates the use of non-human animals in this research as unacceptable. This is because pro-life campaigners objected to creating embryos on the basis that human life begins at conception and invoked human exceptionalist arguments in relation to interspecies entities. This can be contrasted with a similar definition of interspecies embryos as human described in the second discourse – *human and non-human animal nature as materially dynamic*. Here, the stem cell scientists privilege the nuclear genetic material over that provided by the egg (the mitochondrial genome) in their account of interspecies entities as *becoming* human over time (cf. Brown, 2009). While these two discourses deploy the same definition of interspecies entities as human, they have different implications for policy and practice. The former implies that the creation of interspecies entities of all kinds should be banned, while the latter implies a more permissive regulation of such entities – particularly interspecies embryos.

The regulatory implications of how one defines interspecies entities were not lost on the participants in our public event. A number of participants reflected on the relationship between how we might understand or conceptualize interspecies entities and how they are subsequently regulated. In some instances, then, participants self-consciously deployed particular discourses or definitions of different interspecies entities. Notably, it was openly questioned and discussed whether an interspecies embryo, if defined as other than human, would be subject to HFEA regulation and, more specifically, the 14-day limit that stipulates that a human embryo cannot be grown *in vitro* for more than 14 days. Hence, some participants were concerned that, if defined as other than human, an interspecies entity would be allowed to develop into an adult being because the 14-day limit would not apply. It is these concerns that underpin the third discourse – *human and non-human animal natures*. Here, participants implicitly acknowledged the unequal status of human and non-human animals where the latter are more likely to be subject to exploitation. Once viewed in this context, the plural understandings of what constitutes a human and the de-emphasis of ideas of purity can be seen as pragmatic ethical decisions, grounded in their knowledge of contemporary animal-human relations. Hence, regardless of the human or non-human animal material present in an entity, its identity must remain human in order to fall within the remit of the HFEA. And it is this regulatory concern that is considered by some as the primary issue at stake.

At other times in the discussion, however, how we conceptualize these entities is less tied to regulatory concerns but is considered a complex philosophical concern – albeit one with material consequences for human and non-human animals. On the one hand, the third discourse – *human and non-human animal natures* – perhaps presents the greatest challenge to conventional binaries because it offers an understanding of human and non-human natures without recourse to notions of purity. Thus, it opens a space for thinking about species in less bounded ways or even creating a space for a third, *inter* category. Hence, in the context of interspecies entities, the interchangeability of human and non-human animal material makes the humanist agenda difficult to sustain (see Twine, this volume) and invites questioning regarding the possibility of maintaining human/

non-human animal distinctions. On the other hand, a notable feature of this discourse is the process of naturalization that occurs through it (cf. Franklin, Lury and Stacey, 2000). The very notion of a third category – an inter category – sits ill at ease with the ongoing dichotomized thinking in contemporary society. That is, the presence of a third category implicitly unsettles conventional(ized) categories of natural/unnatural or human/non-human. However, what occurs in peoples' accounts is a process where the ontological and epistemological are interwoven to *re*naturalize what interspecies entities *de*naturalized. As illustrated in the second discourse, distinctions between the human and the non-human that are first destabilized through the creation of interspecies entities are then reasserted through material transformations – these entities *become* human because the non-human genome is overwhelmed by human nuclear genome. In this sense, we can say that interspecies entities are not engendering the undoing or the collapse of existing human/non-human animal boundaries or facilitating new orderings of how we understand and what we do with human or non-human animal bodies. Further, according to some arguments set out above not all interspecies entities even involve creating novel things in nature because stem cells derived from interspecies embryos transform into human cells. Yet, by observing this case as classificatory systems in action it becomes clear that '*nature and culture have become increasingly isomorphic while remaining distinct*' (Franklin *et al.*, 2000: 9, original emphasis).[6] Here, it is difficult to disentangle 'nature' from 'technology' and 'science' – the 'natural' from the artificial.

Finally, in returning to Agamben's 'zones of indifference', analysis of the discourses circulating in our public debate suggests that the border between non-human and *some* human animals is blurred. Specifically, we find women's bodies and non-human animals populating a similar space where both are discursively invoked as material resources for scientific and medical research. Indeed, the material interchangeability of women's bodies (where eggs are their proxy) and non-human animal bodies, in stem cell research, symbolically closes the space between them. While it might be pushing the point to claim that women are universally constituted as less than human, by thinking with Agamben we are prompted to consider the ways in which technoscience and medicine continues to pose questions regarding the ongoing expansion of research using women's bodies and non-human animals. In our public event about interspecies entities, the focus on what constitutes an animal or human and the associated regulatory consequences marginalized other fundamental questions about the exploitation of women and non-human animal bodies. Evidently, further work is required to make sense of those who populate the zones of indifference in order to identify and delegitimate oppressive acts against them.

Acknowledgements

A hearty thanks to all who participated in the public event and the project as a whole. I'd also like to thank colleagues involved in the project, particularly those who helped facilitate this event: Stephen Bates, Ann Bruce, Sarah Cunningham-Burley, Wendy Faulkner and Nicola Marks. Conversations held during the meetings held in preparation for this volume proved particularly valuable when developing this chapter. Siân Beynon-Jones, John Dupré, Nina Hallowell and Joseph Murphy provided insightful feedback on earlier drafts, although any remaining failings are my own.

Notes

1 SCNT, more commonly known as cloning, was first used to create Dolly the cloned sheep (Wilmut *et al.*, 2000).

2 ESRC grant number: RES-340-25-0008.

3 Dupré's ideas are not a million miles away from Haraway's concept of 'material-semiotic actors' (1991). Dupré's concept 'promiscuous realism', like Haraway's 'material-semiotic actors', is concerned with navigating a constructivist-realist position. Both are clear to argue that the object of knowledge *exists* – a mouse is not an imaginary entity that is discursively 'made' but has a physical presence in the world – although it is through language that we articulate our world. As Haraway explains, the 'material-semiotic' idea 'is intended to highlight the object of knowledge as an active, meaning-generating axis of the apparatus of bodily production, without *ever* implying immediate presence of such objects or, what is the same thing, their final or unique determination of what can count as object knowledge at a particular historical juncture' (1991: 200, original emphasis).

4 The stem cell scientists were briefed beforehand to answer technical questions if asked directly but not to dominate the discussion with detailed descriptions of the science of stem cells.

5 Indeed, in response to concerns raised by two other participants (one involved in the regulation of stem cell research and one patient group representative) about animal suffering in scientific research, stem cell scientist B gives an impassioned defence of using non-human eggs because it 'replaces' research using human eggs.

6 It is important to note that whilst people's thinking about interspecies entities is structured by such classificatory systems they are not determined by them. As Strathern has argued, 'habits of thought . . . that reproduce themselves in our communications *never reproduce themselves exactly*' (Strathern, 1992: 6, original emphasis). Thus, it is not inevitable that an interspecies entity will be defined as human rather than animal. Rather, how these things are defined is a result of a complex mix of political, social and cultural factors.

References

Agamben, G., (2004), *The Open: Man and Animal*, Kevin Attell (trans.), Stanford: Stanford University Press.

Barnes, B., (1983), 'On the Conventional Character of Knowledge and Cognition' in Knorr-Cetina, K.D. and M. Mulkay (eds), *Science Observed: Perspectives on the Social Study of Science*, London: Sage.

Benton, T. and S. Redfearn, (1996), 'The Politics of Animal Rights – Where is the Left?' *New Left Review*, 215: 43–58.

Bowker, G.C. and S.L. Star, (1999), *Sorting Things Out: Classification and Its Consequences*, London: MIT Press.

Brown, N., (2009), 'Beasting the Embryo: the metrics of humanness in the transpecies embryo debate', *Biosocieties*, 4(2–3): 147–163.

Brown N. and M. Michael, (2001), 'Switching Between Science and Culture in Transpecies Transplantation', *Science, Technology and Human Values*, 26(1): 3–22.
Dupré, J., (1993), *The Disorder of Things: Metaphysical Foundations of the Disunity of Science*, London: Harvard University Press.
Dupré, J., (2002), *Humans and Other Animals*, Oxford: Oxford University Press.
Fox, M., (2005), 'Reconfiguring the Animal/Human Boundary: The Impact of Xeno Technologies', *Liverpool Law Review*, 26: 149–167.
Franklin, S., C. Lury and J. Stacey, (eds), (2000), *Global Nature, Global Culture*, London: Sage.
Haraway, D., (1991), *Simians, Cyborgs and Women: The Reinvention of Nature*, London: Free Association Press.
Haraway, D., (2008), *When Species Meet*, London: University of Minnesota Press.
Holmberg, T., (2005), 'Questioning the "Number of the Beast": Constructions of Humanness in the Human Genome Project (HGP) narrative', *Science as Culture*, 14(1): 23–37.
House of Commons Science and Technology Committee, (2007), Government Proposals for the Regulations of Hybrid and Chimera Embryos, Fifth Report of Session 2006–2007, 05 April 2007, (assessed 25 March 2008) http://www.publications.parliament.uk/pa/cm200607/cmselect/cmsctech/272/272i.pdf
Human Fertilisation and Embryology Authority, (2007), Hybrids and Chimeras: A Consultation on the Ethical and Social Implications of Creating Human/Animal Embryos in Research, London: Crown Copyright.
Marks, J., (2002), *What It Means To Be 98 per cent Chimpanzee: Apes, People and their Genes*, London: University of California Press.
Parry, S., (2009), 'Stem Cell Scientists' Discursive Strategies for Cognitive Authority', *Science as Culture*, 18(1): 89–114.
Strathern, M., (1992), *Reproducing the Future: Anthropology, Kinship and the New Reproductive Technologies*, Manchester: Manchester University Press.
Roberts, C. and K. Throsby, (2007), 'Paid to Share: IVF Patients, Eggs and Stem Cell Research', *Social Science & Medicine*, 66: 159–169.
Wilmut, I., K. Campbell and C. Tudge, (2000), *The Second Creation: The Age of Biological Control by the Scientists Who Created Dolly*, London: Headline.

Part Five
Public Natures

Drawing bright lines: food and the futures of biopharming

Richard Milne

Biotechnology has always been about the future. The synthesis of insulin and somatotropin towards the end of the 1970s prompted a genetic 'gold rush' (Wright, 1994). Established pharmaceutical companies such as Eli Lilly invested in new ones such as Genentech, foreseeing a stream of biotechnological medicines. At around the same time, agronomists envisaged a world transformed by agricultural biotechnology, in which Saudi Arabia would come to resemble the wheat fields of Kansas (Mintz, 1984, in Kloppenburg, 2004).

The future promise of agricultural and pharmaceutical biotechnologies has been accompanied, challenged and contested by less hopeful visions. In Europe at least, debates around genetically modified crops continue, while biotechnological medicines are only now even beginning to realize their potential (Hopkins *et al.*, 2007). Yet this has led, not to a debunking of biotechnology, but to a re-imagining of futures and a new wave of expectations as the new genetic sciences continue to be characterized by their engagement with a 'future yet to be' (Franklin, 2001).

Biopharming represents one such re-imagining of biotechnological promise. It is the production of pharmaceuticals using genetically modified crops, and represents a coming together of medical and agricultural applications of biotechnology. For proponents, the new technology recombines not only plant and human biologies, but also the promises of past applications.

This chapter examines how the promise of biopharming, so-called 'non-food' agriculture (Spök *et al.*, 2008), is performed through the new technology's relations with the cultural and biological materials and meanings of food. I focus on one technical decision, the selection of maize as a platform for biopharming by an EU research consortium, in shaping both hopeful and apprehensive narratives for expert and lay actors. The chapter draws on recent work in STS on technological expectations and 'co-production' (Jasanoff, 2004), which emphasizes that neither 'natural' nor 'social' orders can be taken as causing the other, but that each is produced alongside the other. It explores this relation between existing 'natures' and the imagined futures of biotechnology. It considers how the co-produced biologies and socialities of food support and challenge narratives of the future development of biopharming, the production of pharmaceuticals using genetically modified crops.

Through a range of detailed case studies in areas such as pharmacogenetics, stem cell research and xenotransplantation (Hedgecoe and Martin, 2003; Geesink *et al.*, 2008; Martin *et al.*, 2008; Brown and Michael, 2003) the sociology of expectations has described how the elaboration of hopes and fears for the new biotechnologies is performative, influencing the development of these technologies in the present. Technological promise enrols support for new technologies from publics and funders, and in turn shapes the emergence of new socio-technical assemblages. Expectations form narrative 'scripts' for technology's future development which are 'inscribed' in the artefacts of emerging technoscience (Akrich, 1992; De Laat, 2000). In turn, artefacts come to embody expectations and provide the site at which narratives are encountered by regulators, publics and investors. In a simple form, this can be seen in the 'prototypes' that demonstrate technical viability, or in hubristic performances that reinforce potential and promise, such as drawing an IBM logo on the head of a pin. However, as Jasanoff describes in the case of first generation genetically modified crops, the materials of novel technoscience must take, or make their place among existing products (Jasanoff, 2005a). Expectations are involved not only in the elaboration of new socio-technical configurations, but in the re-working of existing relationships. In turn, both the narratives and materials of new technologies are shaped and re-imagined through their relations with those of existing socio-technical assemblages (Michael, 2006; Konrad, 2006; Rip, 2009). Visions of new technologies tap into socially and culturally embedded imaginaries of science, technology and innovation which are increasingly open to challenge, to re-imagining and to rewriting.

My discussion of biopharming's futures brings together expert and public discussions. It draws on a series of 13 interviews with researchers involved in the EU-based 'Pharma-Planta Consortium'. These are complemented by an analysis of research and regulatory literature published between 2004 and 2008. Together, these sources are used to describe the particular visions for the future of the technology that are being performed within the consortium. They form the majority of the discussion of this paper, but are brought together with focus group research with members of the UK public. In Europe, and particularly in the UK, the GM controversy of the late 1990s acted to expand the constituency of those involved in debates around new technologies.

The interviews were accompanied by focus groups with members of the public. Six groups met in each of London and North West England in early 2007. They met twice, a week apart, and were recruited from local community groups, including sports clubs, groups of work colleagues and a gardening club. The first group meeting explored discussions around foods and pharmaceuticals, while the second built on these using the specific example of biopharming. As biopharming is not a particularly high-profile technology, participants were provided with an information booklet and a link to a website[1] at the end of the first meeting. This booklet featured an introduction to biopharming, and a series of examples of applications that were under active development at the time of the research, including the production of antibodies in maize and tobacco.

By putting these sources alongside each other, I seek to describe how narratives of biopharming emerge around the materialities of 'food'. I chart the increasing complexity of these narratives as they encounter existing sociotechnical assemblages of food agriculture.

An introduction to biopharming

Over the last three decades, a combination of regulation, research funding, industry structures and media 'cultivation' have acted to separate agricultural (green) and biomedical (red) applications (Jasanoff, 2005b; Bauer, 2005). As the 'third generation' of agricultural biotechnology and a development in biopharmaceutical manufacturing, biopharming is both. While first and second generation genetically modified crops focus on food production and consumption, biopharming is a 'non-food' application (Spök *et al.*, 2008), whose goal is the production of pharmaceuticals.

The development of recombinant DNA techniques in the 1970s provided new tools for pharmaceutical research and the manufacturing of pharmaceutical proteins previously extracted from animals (Wright, 1994). Human proteins such as insulin, previously derived from pigs, and purified monoclonal antibodies, used in diagnostic tests and cancer treatments, could be produced on an large scale. The new techniques allowed biological and cultural restrictions on biopharmaceutical availability, such as the size of animals or the availability of pig by-products, to be avoided. Since 1982, when Genentech's recombinant insulin was approved by the FDA, a range of pharmaceutical proteins have been produced in *E. coli* bacteria, the yeast *S. cerevisiae* or in animal cell lines, predominantly Chinese hamster ovary (CHO) cells or baby hamster kidney cells (Walsh, 2007).

The synthesis of insulin was seen 'not merely as launching a new product, but also as beginning a new world' (Bud, 1993: 181). However, biopharming researchers argue that current pharmaceutical production processes involve requirements for labour, equipment and expensive culture media that make the production of only the most valuable proteins economically feasible (Twyman *et al.*, 2005). They argue that biopharming addresses these concerns, provides advantages over existing production, and can meet a rapidly increasing demand for biopharmaceuticals.

Expectations for biopharming centre on cheap, easily scaleable, safe and low-tech pharmaceutical production. These are based on the characteristics of agriculture, as described by Julian Ma and colleagues (2003) in their introduction to the technology:

> Imagine a world in which any protein, either naturally occurring or designed by man, could be produced safely, inexpensively and in almost unlimited quantities using only

> simple nutrients, water and sunlight. This could one day become reality as we learn to harness the power of plants for the production of recombinant proteins on an agricultural scale (Ma *et al.*, 2003: 794).

Biopharming is presented as a pharmaceutical development moving to appropriate the advantages of agricultural production. Since the first experimental demonstration in 1986 (Barta *et al.*, 1986), a number of biopharmaceutical proteins have been expressed in plants, including recombinant proteins, antibodies and vaccines. However, although a vaccine for Newcastle disease in poultry was approved by the US FDA in 2007, no products for humans have yet completed clinical trials. The technology thus continues to exist only in the future, as a potential to be realized, based on 'harnessing the power of plants'.

In this chapter, I focus on the development of biopharming within a large EU Framework Programme Six project, Pharma-Planta. The Pharma-Planta Consortium (PPC) involves over 30 academic laboratories and small biotechnology firms. It was established in 2004 as a collaboration between nearly 40 academic research laboratories and small biotechnology firms in the European Union, with one partner in South Africa. The consortium is led by researchers at St George's Hospital London and at the Fraunhofer Institute for Molecular Biology in Aachen, Germany.

The project's primary goal is to enter a plant-derived pharmaceutical product into clinical trials by the end of the five-year programme. In the process, the consortium aims to achieve other goals – to elaborate a regulatory pathway in the EU, to establish the intellectual property situation, and to lay the foundations for the secondary development of biopharming by European pharmaceutical companies and small biotechnology firms. Given these goals, the number of participants and their prominence within European biopharming research, the PPC represents an influential actor in the elaboration of a vision for the future of the technology.

A key stage in the production of the PPC's narrative for biopharming is the selection of a crop 'platform'. Unlike bacterial and mammalian systems, which have clearly defined 'model systems' in the form of *E.coli* or CHO cells, biopharming's diversity in pharmaceutical targets is accompanied by that in plant species used for their production. Eleven plant species are being used for products currently in clinical trials, with more at the research stage (Basaran and Rodríguez-Cerezo, 2008). This diversity makes it difficult to establish standardized regulatory or intellectual property frameworks, and is a significant hurdle in the wider adoption of biopharming, as it presents 'a bewildering array of choices' (Ramessar *et al.*, 2008a) to pharmaceutical companies. The selection of a production crop thus represents a key point in materializing the futures of biopharming. It embodies the future narratives of biopharming, and provides a point at which these narratives can be encountered, challenged and developed by other actors through their relation with existing socio-technical assemblages.

Of the eleven crops, six can be tentatively classed as non-food crops and five as foods, including rice, potato and maize (Basaran and Rodriguez-Cerezo, 2008). Researchers participating in the Pharma-Planta Consortium (PPC) were involved in developing many of these, but the PPC itself focussed on two crops, maize and tobacco. For the purposes of its 'fast-track' application, which aimed to enter clinical trials by the end of 2009, the PPC chose a monoclonal antibody-based[2] HIV microbicidal cream, produced in genetically modified maize plants (Sparrow *et al.*, 2007).

As I explore in the first part of this chapter, the selection of maize creates not only expectations, but a range of apprehensive futures for actors outside the research community – for NGOs, regulators and food companies. Rather than embodying the promise of biopharming as non-food agriculture, maize's identity as a food crop represents a major challenge. In the second section of the discussion I explore the ways in which researchers draw on a more complex description of the networks of food to make maize central to their narrative of biopharming's future. I then draw on public discussion to consider the further multiplication of complexity and investigate how discursive and material boundaries of 'food' are produced and re-produced in distancing pharmaceutical maize from food.

Pharmaceuticals in maize, drugs in cornflakes

Maize brings together diverse biologies, infrastructures, knowledges, food cultures and social orderings. It was first domesticated in Mexico more than 9000 years ago, and has become a mainstay of global agriculture, ubiquitous in the modern diet in its processed forms.

The Pharma-Planta Consortium is not the first to have picked maize as its preferred production crop. Between 1999 and 2004, 71 per cent of the biopharming field trial applications received by the USDA were for maize (Elbehri, 2005). In Europe, reaction to the production of pharmaceuticals in maize has been critical. Non-governmental organizations such as Greenpeace, Friends of the Earth and Genewatch UK, heavily involved in debates around first generation genetically modified crops, have continued to be critical of the development of biopharming (Mayer, 2003; Allsopp and Cotter, 2005). In a detailed discussion of biopharming, Genewatch link their concerns to the use of food crops, highlighting the risks of gene flow to other crops and the accidental consumption of drug-producing crops (Mayer, 2003). Such concerns are illustrated by the case of Prodigene.

In the 1990s, biotechnology firm Prodigene led the way in developing maize-based pharmaceutical platforms. Their October 2002 agreement with Sigma-Aldrich to produce recombinant trypsin in maize seemed to represent a significant validation of the potential of plant-made production systems. However, almost immediately afterwards, Prodigene was found to have violated the US Plant Protection Act at test plots in Nebraska and Iowa, risking contamination of crops intended for foods. Combined with costs for destroying potentially contaminated crops, the fine imposed by the USDA contributed to the eventual bankruptcy of Prodigene (Spök, 2007).

The Prodigene case brought concerns about the use of maize in biopharming to the fore. It also led to the re-alignment of many of the actors involved in the development of first generation genetically modified crops in the USA. The US food manufacturer's body, the Grocery Manufacturers Association (GMA) had been supportive of the use of first and second generation GM. However, following Prodigene, it argued that:

> there needs to be a presumption against the use of food or feed crops for drug development and manufacture (GMA 2003).

The GMA went on to suggest that a 'bright line' needed to be drawn between commodity food and pharmaceutical farming. As did the concerns of European environmental NGOs, the GMA focussed on the potential for the contamination of the food supply with pharmaceuticals, either by the direct consumption of biopharmed maize or through the cross-pollination of biopharmed and conventional varieties (Mayer, 2003). These concerns were exacerbated by the wind-pollinated biology of maize, and by the ubiquity of the crop in North American agriculture.

Following Prodigene, not only the GMA, but major scientific actors including the journal *Nature* took a strong stance against the use of food crops, supported by the introduction of a 'zero-tolerance' policy on contamination by the USDA (Elbehri, 2005). *Nature* (Anon. 2004) provocatively asked:

> is this really so different from a conventional pharmaceutical or biopharmaceutical manufacturer packaging its pills in candy wrappers or flour bags or storing its compounds or production batches untended outside the perimeter fence? (Anon. 2004: 133).

Nature's concern arises from the juxtaposition in the same form of active, functional pharmaceutical molecules and foods. *Nature* draws on images of pills in sweet packets or in flour to emphasize its point, but others have effectively used a different image to capture concerns about the use of food crops in pharmaceutical production:

> someone is going to get prescription drugs or industrial chemicals in their cornflakes (Larry Bohlen of US Friends of the Earth in Fox, 2003: 4).

The Prodigene experience united actors from opposite sides of earlier debates around GM in opposition to the use of maize. The motif of drugs in cornflakes is a recurrent one in discussion of biopharming, and demonstrates the centrality of 'everyday' food uses of maize to criticisms of biopharming. At the heart of these discussions of biopharming is the 'someone' of Bohlen's comment – the 'at risk' consumer threatened by the contamination of their food supply. These consumers were at the heart of European controversies around agricultural biotechnology, and represent a significant audience for, and participant in, the elaboration of technological futures. I consider these perspectives below, drawing on focus group discussions of biopharming. However, I first consider the reasons of biopharming researchers for the use of maize, in light of the failure of Prodigene and subsequent opposition.

Food-based futures

For opponents of maize-based biopharming, the relationship between pharmaceutical production and food is a simple one. The risk of contamination precludes the use of the same crop for non-food purposes. By placing maize at the heart of their narrative of the future of biopharming, the Pharma-Planta consortium appear to risk undermining this future in the face of sceptical publics and outright opposition from influential actors in both the United States and Europe. However, biopharming's promise is based on its agricultural qualities. In interviews, and in a series of papers (Sparrow *et al.*, 2007; Ramessar *et al.*, 2008a; Ramessar *et al.*, 2008b), PPC-affiliated researchers describe the basis for selecting maize. They suggest that maize rather than tobacco represents the best option for the future development of the technology. In particular, it is the qualities of maize as a food that separate it from tobacco as the preferred production platform. The diversity of these qualities emphasizes the heterogeneous identity of a 'food' crop, and aligns an array of biological, social and technological qualities of the crop.

The importance of maize in agriculture has made it a central focus of scientific research and technical innovation. The selection of maize as a pharmaceutical production platform reflects this accumulated weight of scientific knowledge and expertise around a particular plant biology. This is the initial position adopted by PPC researcher Koreen Ramessar and colleagues in a defence of maize-based production, which sets out to establish the advantage of maize and to:

> dispel the misconception that it is better to use less known and non-domesticated plants . . . instead of a well-understood and familiar species (Ramessar *et al.*, 2008b: 410)

The advantage of using food crops, and particularly maize are asserted through the concentration of knowledge around food. As they go on to suggest the choice of maize represents the self-reinforcing weight of tradition – a technological 'lock-in' (Stirling, 2008) created by the co-production of scientific knowledge and agricultural practice:

> The focus on maize has led to more information accumulating in this species than any other about mechanisms to maximize transgene expression and product recovery (Ramessar *et al.*, 2008b: 412).

The scientific knowledge that has 'accumulated' around maize in the past contributes to shaping the researchers' vision for the future of biopharming. The characteristics of maize that render it attractive are those of its biology, and the relationship between this and the wider assemblage of agriculture and food production.

Food biologies

Ramessar and colleagues (2008b: 409) describe maize as 'an ideal production platform for safe and effective molecular farming'. A tradition of agricultural research orientates this focus, suggesting a value in continuity from foods through to pharmaceutical production. This narrative is supported by a range of qualities of maize biology.

Firstly, recombinant proteins produced within maize are stored within the seeds. The advantages of this are described in an interview with a biopharming research professor:

> the seed . . . is a tissue which is biologically developed for protein storage, so you get tonnes of maize seed which is concentrated antibody and desiccated, and pre-packaged for you if you like, and you can leave that in a silo for ten years if you like before you process it (Professor L.)

Maize is not simply a production vehicle, but is integral to expectations of biopharming. Its stable, desiccated seeds provide food crop biopharming with a 'natural' advantage over cell cultures, which must be maintained, and over non-food crops such as tobacco, whose leaves must be processed for protein extraction. Maize is also able to fold these proteins and undertake post-translation modifications (such as the addition of sugars) in ways that cannot be done by other plants or bacterial systems.

Maize's biological qualities are themselves the product of centuries of interaction between people and food, as the interview extract below emphasises:

> if you look at the biology of different crops, you get the far best yields from a crops that has been bred for food . . . You know people talk about well, why don't we just find another plant that's not used for food and domesticate that? But domestication takes hundreds and hundreds of years, and to breed in all the characters you needed . . . it is not a trivial job (Professor C.).

Prior to and since the development of agricultural biotechnology, maize's biologies have proven more amenable to transformation than other cereal varieties (Ramessar *et al.*, 2008b). They do not represent an intrinsic, pristine 'nature' that lends itself to pharmaceutical production, but rather are co-productions, the embodiment of work already done that establishes the potential for future progress. The emphasis on the co-produced biologies of maize acts to reinforce the value of continuity, and to rule out alternative visions based on the use of non-food crops.

Food infrastructures

The biologies of maize exist within a wider assemblage of food production. These contribute as much as the crop itself to establishing expectations for food-crop based pharmaceutical production. As described in an interview with a molecular biology professor:

> . . . people choose edible plants because they're not toxic, they have no toxins, and an awful lot is known about processing, because they're actually used (Professor A.).

The existing uses of maize are crucial to research expectations for the technology. Food uses of the crop have established a large body of expertise in the farming of maize and the processing of maize kernels. The co-option of existing farming practices and expertise for pharmaceutical production is complemented by the availability of food-based extraction techniques. As Ramessar and colleagues continue:

> all pharmaceutical maize crops need to go through similar grinding and extraction steps early in the process to convert the seeds into powdered meal and release the target proteins . . . milling is a very well established and inexpensive process in the food industry (Ramessar *et al.*, 2008b: 413)

In one laboratory, I was shown the domestic food processing equipment used for the initial processing of the crop. The expectation here is that this will be scaled up to make use of the industrial equivalents. For the PPC, the use of existing infrastructure also extends the promise of biopharming beyond the regions that dominate conventional pharmaceutical production to anywhere that has processing capacity for plant material. The ubiquity of processing technologies reinforces the selection of maize as a pharmaceutical platform. It suggests that the low cost of biopharming can be achieved through the production of plant made pharmaceuticals using existing tools and facilities.

Food safety

The knowledges, infrastructure and biologies described above come together in researcher discussions of maize's role in establishing the safety of plant made pharmaceuticals. For biological medicines, 'the process is the product' (Longstaff *et al.*, 2009). The control and standardization of the production process is paramount in establishing the quality and safety of the biopharmaceutical product. In the case of existing manufacture, quality and purity regimes have developed around the particular biologies of *E. coli* or Chinese hamster ovary cells (Walsh, 2007). These platforms are linked to elaborate processes of purification and quality control, complicated in the case of mammalian cells by the potential for contamination with zoonotic diseases. In contrast, plants are described as inherently safe:

> no plant pathogens are known to infect humans – after all, most people eat fresh or cooked plant material every day with no ill effects (Twyman *et al.*, 2005: 2).

The inherent safety of maize, it is claimed, represents a key advantage over production using mammalian cell cultures. That maize is a plant means that it is inherently unlikely to contaminate the pharmaceutical product. Equally importantly, its status as a *food* plant means that it becomes possible to take short-cuts in establishing the safety of the product. Unlike alternative platforms such as tobacco, maize, as a food, is inherently non-toxic. Again, food biologies provide an advantage over non-food. In a review of the regulations for bio-pharming in Europe and the USA, Sparrow *et al.* (2007) describe how an understanding of the safety of foods is performed in regulatory understandings. In particular, they suggest the relevance of regulatory classifications that establish food safety on the basis of 'a history of safe use'.

The assertion of safety on the basis of prior use is familiar from debates around first generation genetically modified crops (Levidow *et al.*, 2007). Establishing the similarity of the new technology to existing food production techniques was central to the 'ontological ordering' (Jasanoff, 2005b: 147) of biotechnology. The adoption of a 'product' approach to regulation positioned the novel ontologies of biotechnology 'on the side of the familiar and manageable' rather than 'on the side of the unknown and insupportably risky' (Jasanoff, 2005b: 139).

The use of classifications such as 'substantially similar' or 'substantially equivalent' has inserted the risks of the new products into a relationship with their 'conventional', 'traditional', or 'old established' (OECD, 1993) counterparts. The definition of biotechnology in these terms contributed to black-boxing it as a precise, predictable and incremental innovation indistinguishable from conventional agricultural advance (Plein, 1991). The establishment of genetically modified varieties as 'as safe' as traditional or conventional foods relies on the safety of these themselves. The regulation of foodstuffs in both the UK and the USA draws on the biologies, knowledges and cultures of food to establish this. Traditional foods are classed as having a 'history of safe use' (in the EU), or as being 'Generally Recognized as Safe' (GRAS) (USA). Establishing the safety of traditional varieties allows them to act as a comparator for genetically modified varieties. In doing so, the relevant biologies of food crops must be established and tested. In the late 1990s, the application of such approaches to the safety testing of genetically modified crops was heavily criticized on the basis that it unreasonably emphasized chemical equivalence (Millstone *et al.*, 1999).

OECD 'consensus' documents bring together expert knowledges to clarify and define the relevant 'biologies' of crops such as maize for use in comparison (OECD, 2003). However, in both Europe and the USA, decisions which establish the continuity of novel foods with traditional varieties are relevant only when they are prepared, processed and used 'under the condition of intended use' (Burdock and Carabin, 2004: 5; Constable *et al.*, 2007). These safety designations acknowledge that the biologies of traditional food are often potentially toxic yet are rendered safe by with traditional preparation practices.

The establishment of equivalence is thus an outcome of food biologies and cultures. As biopharming inserts potentially active molecules into crops, and involves a change in intended use, it appears unlikely that pharmaceutical-producing maize would easily be established as food. Nevertheless, for biopharming researchers, maize's classification as GRAS plays an important role in asserting its pharmaceutical safety:

> maize as a food crop has FDA GRAS status which provides an advantage over non-food crops and other plant systems. The data for allergenicity, toxicity and dietary exposure to maize is already available, as is data on worker exposure and safe handling . . . comparing the pharmaceutical crop to existing data is sufficient (Ramessar *et al.*, 2008b: 416)

The traditions and biologies of crop plants, which combine to establish the safety of novel food products, are enrolled into the pharmaceutical promise of biopharming. They provide the technology with a key advantage over both non-food crops and existing biopharmaceutical manufacturing. However, the advantages of maize derive from its position at the heart of concerns about food safety and genetically modified agriculture. In the following section, I explore how this association with food, rather than supporting researcher expectations of the technology, represents a challenge to realizing them.

Increasing complexities, unstable boundaries

For those developing plant made pharmaceuticals, maize represents an apparently logical choice as a production platform. The intrinsic qualities of the crop, and its location within a tradition of agricultural research and wider networks of food production and processing, contribute to establishing high expectations for the technology. Yet by incorporating the heterogeneous qualities of food agriculture into biopharming's future, researchers enter the technology into a set of complex relations, not all of which are equally supportive of this future. In light of these, the classification of biopharming as 'non-food' agriculture seems increasingly unstable, as highlighted in the critical reaction of actors such as the GMA and *Nature*.

In this section I consider how these boundaries between food and non-food are re-negotiated as researchers attempt to re-establish the promise of the technology. In doing so, I bring the discussions of researchers and other actors introduced to date together with that of publics as 'everyday' consumers of food. As described earlier, the imagination of technological futures is no longer the preserve of researchers and governments alone. Indeed, the views of publics are described as a critical 'second hurdle' in the future of biopharming by researchers (Ma *et al.*, 2005). The relationship between biopharming and foods was explored in focus group discussions in which participants considered the use of maize and other food crops for pharmaceutical production.

As the biologies, cultures and knowledges of food come to support and challenge its suitability for pharmaceutical production, it is not only the characteristics of food that are mobilized by expert and lay actors, but the distinction between food and non-food, as they attempt to construct the inedibility of plant made pharmaceuticals.

Preserving the identity of food

NGOs and critical scientific opinions have opposed the production of biopharmaceuticals in maize, citing the risk of contamination of the environment and more particularly of the food supply. These criticisms focus on the role of maize as human food, symbolized by cornflakes. In focus group discussions also, concern emerges around the contamination of food supplies. While participants did not explicitly discuss the Prodigene case, some of their concerns echo those of the GMA:

> Rita: I think the question of contamination is . . . the biggest consideration there is [. . .] It might be small quantities at the beginning, and then the quantities keep increasing over time. And then that is one of, one of the real dangers that will creep in (Young urban female, meeting 2)

As Rita describes, a main concern about biopharming is the contamination of the food supply. However, rather than the image of finding drugs in cornflakes, Rita's concerns are of a more insidious or gradual 'pollution' which builds up to hazardous levels. Pharmaceutical contamination is presented in a similar way to that by pesticides and agricutural chemicals, but also echoes the framing of first generation GM in terms of 'genetic pollution' (see Levidow, 2000).

A first response to preventing the contamination that maize production threatens consists of reinforcing a boundary between food and non-food agriculture. In particular, this involves establishing (or re-establishing) a visible and/or material boundary between food and non-food products. The majority of means proposed, including containment and the adoption of a parallel system of processing, lose the infrastructural and scalar qualities of maize that make it part of biopharming's future. Consequently, I focus on efforts to establish a boundary between food and non-food maize products by means of 'identity preservation', by establishing material, visible differences between food and pharmaceutical products. The use of specific genetic constructs, and/or the use of particular varieties of plants allows the pharmaceutical product to be visibly placed at all times, as Professor B. describes:

> You can use seed specific promoters, so you know your product is not going to be anywhere else in the plant, and you can use a coloured strain of maize . . . so you can make sure that all the seed that's got your drug in it is pink. So any cross contamination of the food chain could be immediately and easily seen (Professor B.).

In a report for the UK Department of Food and Rural Affairs on the containment of genetically modified crops, Dunwell (2005) adds the possibility of using green and purple tomato varieties. These methods again demonstrate the co-production of the 'biological' and 'social' qualities of food in the futures of biopharming. The use of coloured variety provides a biological basis on which to establish boundaries between food and non-food. However, as is the edibility of maize itself, the inedibility of these (biologically edible) varieties is a result of food cultures within which the consumption of red tomatoes or yellow maize is normalized.

In the discussion below, participants address the technical content of identity preservation, which was presented in the information booklet provided. Their discussion highlights the extent to which both the biological and cultural characteristics of food form part of the wider relations into which biopharming enters. An initial challenge to the ability of the technique to effectively separate foods and pharmaceuticals is presented by participant understandings of plant biology:

> Evie: How do you call that word when cells change?
>
> Sarah: Mutate
>
> Evie: Yeah, you don't know what'll happen, even if you do coloured varieties or maybe infertile varieties become fertile, so you don't know
> (Young urban female, meeting 2)

The mutability of cells contributes to scepticism about the visibility of biopharmed products. The ability to separate pharmaceutical non-food from food consistently is threatened by an inherent uncertainty of plant genomes. In addition, focus group discussion suggests that the separation of pharmaceutical and food products is reliant on the imagination of particular future consumers:

> Hilda: That thing there, using genes to make the crop look distinctive, now, I don't think that's bad, a white tomato or a red one, you do know then, if you planted everything and they all looked the same and they weren't the same, how would you know? I think this is what people are worried about. The little bit that I read or heard a long time ago about all this GM coming, you know, to the state that it's in now, I think people were worried about that, because they didn't know what they were getting and what was in it,
>
> RM: So not being able to tell?
>
> Hilda: You have a choice then don't you, you've no choice if you don't know, do you?
>
> Elsie: I think definitely making the crops look distinctive, they'd have to look completely different, because somebody would go and use the wrong ones wouldn't they in this country, definitely, they would
> (Older rural female, meeting 2)

The use of unconventional varieties potentially allows participants to reassert their ability to choose between foods, and reduces concern about the use of food crops. However, the effectiveness of separating pharmaceutical crops from food relies on not only distinctive crop biologies, but on the behaviour of consumers. While allowing consumers to distinguish between food and pharmaceutical products is important, it is also necessary to ensure sufficient distinctiveness. The potential effectiveness of identity preservation is challenged by the likely problems with consumers 'in this country'. Similarly another group problematizes the role of publics in identity preservation:

> Mark: What are coloured varieties?
>
> RM: Well, for example, if you modified a white tomato, you'd be able to see which ones were modified
>
> James: I don't think that would work as a good method of containment, say you create blue maize, someone comes along and says oh my god, there's one strand of blue maize in my crop field, then you have nationwide panic of this maize is spreading
> (Young urban male, meeting 2)

James suggests that the social reception of identity preserved varieties will not be that intended by researchers, and may result in 'nationwide panic'. Drawing on the example of an outbreak of H5N1 avian influenza in the UK in early 2007, he suggests that this would result in 'irrational' fear from other publics. His concerns reveal the extent to which identity preservation assume particular consumer and regulatory responses. While no comment can be made here on the technical certainty of identity preservation as a means for separating foods and non-foods, it has only limited potential as a means of separating hopeful pharmaceutical futures from food-based apprehensions.

What is a food?

The discussion of identity preservation concentrates on an intervention whose primary aim is to materially separate foods from non-foods. The proposed solutions of researchers are challenged and problematized by consumer narratives of the unpredictability of crop biologies and consumer behaviour. A second approach to establishing a boundary between maize-as-food and maize-as-drug focuses on the very classification of crops as food.

In introducing researcher narratives of biopharming's potential, I described how an accumulation of research expertise contributes to the selection of maize. This research expertise has been produced together with the food biologies of the maize crop through centuries of interaction. However, in the interview extract below Professor L. disrupts this relationship to re-localize the production and consumption of maize. He suggests that the food status of maize is itself contestable and is situated in local food cultures, rather than a universal attribute of the crop.

> A food in one country is not necessarily a food in another country. So in the UK we don't really grow maize as food, the conditions aren't right, you know, we grow

wheat, barley whatever. You know, you'll occasionally come across a maize field that is grown for animal feed. We don't grow maize basically. So I don't regard maize in England as a food product. So why shouldn't we grow pharmaceutical maize? (Professor L.)

The problems of maize production are situated in particular cultural geographies of consumption, which distinguish them from an imaginary of British or English food production. Indeed, maize farming is rare in the UK, where it is grown mainly as forage.

While Professor L. draws on food cultures to distinguish maize from British food production and consumption, public discussion describes the opposite. In the earlier group discussion of biopharming, Rita emphasised that the main threat of biopharming was the contamination of food crops. However, in the extract below, Rob and Mark describe how the complexity of food production makes it difficult to assess the impacts of food crop biopharming. In doing so they bring the heterogeneity of food networks back to the fore, and introduce new actors. Their discussions open up discussion of maize from a two-way relationship between crop and humans to incorporate non-human biologies and behaviours:

Rob: I'd like to know what would happen to the animals that eat the maize, like field mice and things like that, and the insects that feed on them, like how it would impact on the food web and sort of things like that, I don't know maybe that. . .before it was used widely

Mark: But that's part of the problem, it's things like that that are really hard to find out, isn't it, because it's so complex'
(Young urban male, meeting 2)

As does Professor L., Rob and Mark extend understandings of what 'food' means by asking 'food for whom?' However, while Professor L.'s description narrows the boundaries of food so as to exclude maize, that from the focus group expands these boundaries, and suggests that the use of food crops involves a set of interrelations with animal as well as human foods. Moreover, once food networks are extended, the separation of food from non-food crops itself becomes problematic, as a participant in the same group considered in discussing tobacco:

Ed: Why is tobacco not, did I read this wrong? Why is tobacco not in the food chain? Does nothing eat tobacco? Not even insects?'
(Young urban male, meeting 2)

A participant in another group continued this challenge to the distinction between food and non-food introduced in the case of tobacco:

Jane: On your second page here, where it's on about tobacco, it actually says tobacco is not used for food or animal feed so its unlikely it would contaminate either the human or animal food chain, but it could, through the predators on the plants, so I'm not sure why that statement's been made really (Older rural mixed, meeting 2)

For Jane, the distinction between food and non-food agriculture is meaningless. In these terms, tobacco and maize become part of one and the same all-encompassing food network. In emphasising the biological continuity of food, focus group participants contest the separations performed by Professor L. These latter extracts suggest that as discussions of the biologies and socialities of crops become increasingly well-developed, the number of human and non-human participants proliferates. Consequently the separation of food from non-food in the development of biopharming becomes increasingly problematic.

Food and unstable promises

Biopharming is among the latest in a long line of promising biotechnological innovations. Although researchers have succeeded in demonstrating a 'proof of principle', there remains a long road to the commercialization of plant derived pharmaceutical crops. A key step on this road is the elaboration of a standardized production platform. However, the choice of a crop for production represents the materialization of particular narratives for biopharming, and the closing down of other technological options. It enters the technology into a set of socio-technical relations. Through the example of Pharma-Planta's use of maize, I have described how the use of a 'food' crop both underpins and undermines the promise of the technology.

The artefacts of novel technoscience are the point at which visions of their future are materialized and encountered. In the case of biopharming, the materialities of maize become the point around which these visions become controversial as relations between pharmaceutical 'non-food' agriculture and foods are established and contested. For some, maize's centrality to the failures of Prodigene demonstrates biopharming's threat to the futures of safe consumption. Disparate actors from contrasting sides of the GM debate became allied in opposition to the use of food crops, and maize comes to embody fears about the technology's future. Yet for researchers, the same plant embodies the advantages of the safe, easily purified biologies of the plant material which are supported by extensive infrastructure.

The material form of new technologies does not necessarily represent the stabilization of expectations, nor the finalized form in which expectations are encountered. The role of maize in biopharming highlights the ways in which the same materiality can come to embody both hopes and fears around new technologies as the complexities around it multiply. The imagined materialities of biopharming are incipient in the realization of the technology's futures. They mutate and evolve as more actors become involved in the creation of new socio-technical networks around biopharming and the restructuring of those around food.

Notes

1 http://www.homepages.ucl.ac.uk/~ucrhrjm/research/mf.html
2 Antibodies bind specifically to antigens on the surface of cells and allow highly accurate diagnostic tests and the specific targeting of drug treatments. Monoclonal antibodies represent the largest class of biopharmaceuticals in development (Walsh, 2007).

References

Akrich, M., (1992), 'The description of technological objects', in W.E. Bijker and J. Law (eds), *Shaping Technology/Building Society: Studies in Sociotechnical Change*. Cambridge, MA: MIT Press: 205–224.

Allsopp, M. and J. Cotter, (2005), *Pharm crops: a super-disaster in the making*, Exeter, UK: Greenpeace Technical Report.

Anon., (2004), Drugs in crops – the unpalatable truth. *Nat Biotech*, 22(2): 133–133.

Barta, A., K. Sommergruber, D. Thompson, K. Hartmuth, M. Matzke and A. Matzke, (1986), 'The expression of a nopaline synthase – human growth hormone chimeric gene in transformed tobacco and sunflower callus tissue', *Plant Molecular Biology*, 6(5): 347–357.

Basaran, P. and E. Rodríguez-Cerezo, (2008), 'Plant Molecular Farming: Opportunities and Challenges', *Critical Reviews in Biotechnology*, 28(3): 153–172.

Bauer, M., (2005), 'Distinguishing Red and Green Biotechnology: Cultivation Effects of the Elite Press', *International Journal of Public Opinion Research*, 17(1): 89, 63.

Brown, N., (2003), 'Hope Against Hype – Accountability in Biopasts, Presents and Futures', *Science Studies*, 16(2): 3–21.

Brown, N. and M. Michael, (2003), 'A Sociology of Expectations: Retrospecting Prospects and Prospecting Retrospects', *Technology Analysis and Strategic Managment*, 15(1), 18: 3.

Bud, R., (1993), *The Uses of Life: A History of Biotechnology*, Cambridge: Cambridge University Press.

Burdock, G.A. and I.G. Carabin, (2004), 'Generally recognized as safe (GRAS): history and description', *Toxicology letters*, 150(1): 3–18.

Constable, A., D. Jonas, A. Cockburn, A. Davi, G. Edwards, P. Hepburn, C. Herouet-Guicheney, M. Knowles, B. Moseley, R. Oberdörfer and F. Samuels, (2007), 'History of safe use as applied to the safety assessment of novel foods and foods derived from genetically modified organisms', *Food and Chemical Toxicology*, 45(12): 2513–2525.

De Laat, B., (2000), 'Scripts for the Future: Using Innovation Studies to Design Foresight Tools', in Brown, N., B. Rappert and A. Webster, (eds), *Contested Futures: A Sociology of Prospective Techno-science*. Aldershot: Ashgate: 175.

Dunwell, J.M., (2005), *Technologies for biological containment of GM and non-GM crops. Defra contract CPEC 47*, London: Department for Environment, Food and Rural Affairs (Defra).

Elbehri, A., (2005), 'Biopharming and the Food System: Examining the Potential Benefits and Risks' *AgBioForum*, 8(1), available at: http://www.agbioforum.org/v8n1/v8n1a03-elbehri.htm. (last accessed 14 May 2009).

Fox, J.L., (2003), 'Puzzling industry response to ProdiGene fiasco', *Nat Biotech*, 21(1): 3–4.

Franklin, S., (2001), 'Culturing biology: cell lines for the second millennium', *Health*, 5(3): 335–354.

Geesink, I., B. Prainsack and S. Franklin, (2008), 'Stem Cell Stories 1998–2008', *Science as Culture*, 17(1): 1–11.

GMA, (2003), Food Industry Comments on Proposed FDA Regulations for Plant-Made Pharmaceuticals Feb 6, 2003, available at: http://www.gmabrands.com/publicpolicy/docs/Comment.cfm?DocID=1068& [Accessed July 3, 2009].

Hedgecoe, A. and P. Martin, (2003), 'The Drugs Don't Work: Expectations and the Shaping of Pharmacogenetics', *Social Studies of Science*, 33(3): 327–364.

Hopkins, M.M., P.A. Martin, P. Nightingale, A. Kraft and S. Mahdi, (2007), 'The myth of the biotech revolution: An assessment of technological, clinical and organisational change', *Research Policy*, 36(4): 566–589.

Jasanoff, S. (ed.), (2004), *States of Knowledge: The Co-Production of Science and Social Order*, London: Routledge.

Jasanoff, S., (2005b), 'In the democracies of DNA: ontological uncertainty and political order in three states', *New Genetics and Society*, 24(2): 139–156.

Jasanoff, S., (2005a), *Designs on Nature: Science and Democracy in Europe and the United States*, Princeton: Princeton University Press.

Jasanoff, S., (2006), 'Biotechnology and Empire: The Global Power of Seeds and Science', *Osiris*, 21(1): 273–292.

Kloppenburg, J., (2004), *First the Seed: The Political Economy of Plant Biotechnology (Science and Technology in Society)* 2nd edn., Madison WI:University of Wisconsin Press.

Konrad, K., (2006), 'Shifting but forceful expectations: structuring through the prospect of materialisation', Paper presented at the Twente V workshop Material Narrative – of Technology in Society; Enschede October 19–21, 2006.

Levidow, L., (2000), 'Pollution Metaphors in the UK Biotechnology Controversy', *Science as Culture*, 9(3): 325–351.

Levidow, L., J. Murphy and S. Carr, (2007), 'Recasting "Substantial Equivalence":Transatlantic Governance of GM Food', *Science Technology Human Values*, 32(1): 26–64.

Longstaff, C. *et al.*, (2009), 'How do we assure the quality of biological medicines?' *Drug Discovery Today*, 14(1–2): 50–55.

Ma, J., P. Drake and P. Christou, (2003), 'The production of recombinant pharmaceutical proteins in plants', *Nat Rev Genet*, 4(10): 794–805.

Ma, J.K., E. Barros, R. Bock, P. Christou, P.J. Dale, P.J. Dix, R. Fischer, J. Irwin, R. Mahoney, M. Pezzotti, S. Schillberg, P. Sparrow, E. Stoger and R.M. Twyman, (2005), 'Molecular farming for new drugs and vaccines', *EMBO Rep*, 6(7): 593–599.

Martin, P., N. Brown and A. Kraft, (2008), 'From Bedside to Bench? Communities of Promise, Translational Research and the Making of Blood Stem Cells', *Science as Culture*, 17(1): 29–41.

Mayer, S., (2003), *Non-food GM crops: new dawn or false hope? Part 1: Drug Production*, Buxton: GeneWatch UK.

Michael, M., (2006), *Technoscience and Everyday Life: The Complex Simplicities of the Mundane*, Maidenhead, UK: Open University Press.

Millstone, E., E. Brunner and S. Mayer, (1999), 'Beyond "substantial equivalence"', *Nature*, 401(6753): 525–526.

OECD, (2003), *Consensus Document on the Biology of Zea mays subsp. mays (Maize)*, Paris: OECD.

OECD, (1993), *Safety Evaluation of Foods Derived by Modern Biotechnology: Concepts and Principles*, Paris: Organisation for Economic Cooperation and Development, available at: http://www.oecd.org/dataoecd/57/3/1946129.pdf. (last accessed 30 March 2010).

Plein, C., (1991), 'Popularizing Biotechnology: The Influence of Issue Definition', *Science, Technology, and Human Values*, 16:490: 474.

Ramessar, K., T. Capell and P. Christou, (2008a), 'Molecular pharming in cereal crops', *Phytochemistry Reviews*, 7(3): 579–592.

Ramessar, K., M. Sabalza, T. Capell and P. Christou, (2008b), 'Maize plants: An ideal production platform for effective and safe molecular pharming', *Plant Science*, 174(4): 409–419.

Rip, A., (2009), 'Technology as prospective ontology', *Synthese* 168(3): 405–422.

Roosevelt, M., (2003), 'Cures On the Cob', *Time*, available at: http://www.time.com/time/magazine/article/0,9171,452804–1,00.html [Accessed May 11, 2009].

Sparrow, P., J. Irwin, P. Dale, R. Twyman and J. Ma, (2007), 'Pharma-planta: road testing the developing regulatory guidelines for plant-made pharmaceuticals', *Transgenic Research*, 16(2): 147–161.

Spök, A., (2007), Molecular farming on the rise – GMO regulators still walking a tightrope. *Trends in Biotechnology*, 25(2), 74–82.

Spök, A., R.M. Twyman, R. Fischer, J.K. Ma and P.A. Sparrow, (2008), 'Evolution of a regulatory framework for pharmaceuticals derived from genetically modified plants', *Trends in Biotechnology*, 26(9): 506–517.

Stirling, A., (2008), '"Opening Up" and "Closing Down": Power, Participation, and Pluralism in the Social Appraisal of Technology', *Science Technology Human Values*, 33(2): 262–294.

Strategy Unit, (2003), *Field Work: Weighing up the Costs and Benefits of GM Crops*, London: Cabinet Office, available at: http://www.cabinetoffice.gov.uk/strategy/work_areas/gm_crops.aspx (last accessed 14 May 2009).

Twyman, R.M., S. Schillberg and R. Fischer, (2005), 'Transgenic plants in the biopharmaceutical market', *Expert Opinion on Emerging Drugs*, 10(1): 185–218.

Van Lente, H., (2000), 'From promises to requirement', in Brown, N., B. Rappert and A. Webster (eds), *Contested Futures: A Sociology of Prospective Techno-science*, Aldershot: Ashgate.

Walsh, G., (2007), *Pharmaceutical biotechnology: Concepts and Applications*, London: Wiley-Blackwell.

Wright, S., (1994), *Molecular Politics: Developing American and British Regulatory Policy for Genetic Engineering, 1972–82*, Chicago, IL: University of Chicago Press.

Barcoding nature: strategic naturalization as innovatory practice in the genomic ordering of things

Claire Waterton

> A class is a group of things; and things do not present themselves to observation grouped in such a way. We may well perceive, more or less vaguely, their resemblances. But the simple fact of these resemblances is not enough to explain how we are led to group things which thus resemble each other, to bring them together in a sort of ideal sphere, enclosed by definite limits, which we call a class, a species, etc. We have no justification for supposing that our mind bears within it at birth, completely formed, the prototype of this elementary framework of all classification. Certainly, the word can help us to give a greater unity and consistency to the assemblage thus formed; but though the word is a means of realizing this grouping the better once its possibility has been conceived, it could not by itself suggest the idea of it. From another angle, to classify is to not only to form groups; it means arranging these groups according to particular relations. We imagine them as co-ordinated, or subordinate one to the other, we say that some (the species) are included in others (the genera), that the former are subsumed under the latter. There are some which are dominant, others which are dominated, still others which are independent of each other. Every classification implies a hierarchical order for which neither the tangible world nor our mind gives us the model. We therefore have reason to ask where it was found (Durkheim and Mauss, 1970 [1903]: 8).

Introduction

The quotation above is an excerpt from Durkheim and Mauss' essay entitled 'On Some Primitive Forms of Classification: Contribution to the Study of Collective Representations' (first published in *Année Sociologique* in French in 1903). As both the excerpt and the title of the essay suggest, one of the main concerns (and achievements) of these authors was to move discussion about classifications, and the human propensity to order and classify, away from the idea of there being a pre-existing logic in nature, and away from the idea of an innate human ability to classify things. Based on a critique of the tautological reasoning of the metaphysics and psychology of their day,[1] Durkheim and Mauss, were amongst the first to argue that *sociological* questions about the way in which we order and classify our world are important if we want to understand, first, how classifications are made, and second, what they do. So first, through the use of illustrative examples, they argued that classification is a collective and not an individual endeavour, and not one that is informed solely by nature. Even abstract ideas about time and space are argued to be 'closely connected with [the] corresponding social organization' (1970 [1903]: 88); Selected Australian classifications, to give an example, are described as being consistent with a sense of universal belonging within which no boundaries exist between animate and inanimate parts (1970 [1903]: 18). Not only classes, but the relations between them, according to Durkheim and Mauss, are 'of social origin' and perform social roles (1970 [1903]: 83).

Durkheim and Mauss' additional emphasis on what social work classifications 'do', suggests clearly that, at the beginning of the last century, anthropologists were already making the move from thinking about epistemological questions concerning the 'truth' of classes (their basis in nature, so to speak) to more ontological questions about how classifications and their resultant categories create and sustain social relations. So for example, the complex Chinese classification of eight spatial parts, each connected to eight different powers, each connected in turn to a whole collection of things, is interpreted as a system 'intended above all to regulate the conduct of men' (1970 [1903]: 83). By paying attention to what classifications 'do', Durkheim and Mauss opened up a space for what would turn out to be a longstanding sociological/anthropological fascination with the way in which classifications and the entities they circumscribe perform in the world, what happens through them and in relation to them, what their 'consequences' and effects are (Bowker and Starr, 1999).

Durkheim and Mauss' work is not without its critics,[2] but I refer to it here as a foundation, within social theory, for *thinking sociologically/anthropologically about the collective construction of categories and about what such categorization does*. I highlight this theme at the beginning of this chapter, as a thread that I hope to weave through it, although the kind of ordering of categories that I am about to introduce – the classification and identification of natural organisms – differs substantially from those 'primitive classifications' studied by Durkheim and Mauss, and indeed, from the work of subsequent sociologists and anthropologists who would take up the intellectual challenges they posed.[3]

This chapter examines 'DNA barcoding' or simply 'barcoding' – a new genomic technology for recognizing and ordering species. Barcoding is a recent innovation within the science and practice of taxonomy (the science that is concerned with ordering plants, animals and micro-organisms into groups based on their shared characteristics). It is, essentially, an innovation in method, involving the standardized use of a short gene sequence, taken from a replicable position in the mitochondrial genome of selected natural specimens, to identify such organisms to species level. Treated sociologically/anthropologically, however, barcoding cannot be reduced to mere methodological innovation. This new technoscientific/genomic technique for ordering natural kinds harbours – in monadological fashion – broader socio-political orderings, and it also does particular kinds of work. Like many methodological innovations it both requires and supports an entire socio-technical 'package' (Clarke and Fujimura, 1992), the basic features of which I outline in the following section.

But before looking at what barcoding entails and requires in more depth, I want to return briefly to the passage above by Durkheim and Mauss and to highlight why I think that the question that preoccupied these authors at the beginning of the last century still grips us today, albeit in a very different era of genomic 're-cataloguing', 'rewriting' and the attendant re-configuring of human-nature relationships (Rabinow, 1996; Rose and Novas, 2004; Rose, 2007; Sunder, 2006; Franklin, 2003). The question that pre-occupied Durkheim and Mauss was this: if neither nature nor our innate mental capacities provide an underpinning to the orderings of nature we create, where are these boundaries and the names given to entities to be found? This is a very tense question, it seems to me, at the beginning of the 'biotech century', and is one that is found at the heart of many of the fierce debates surrounding new genomic technologies, especially as they relate to the life and biological sciences. That the question is loaded in the present context, highlights, I would suggest, the enormous weight of our investment in the natural, as well as our understanding of its power and authority. As Daston and Vidal would have it, even as novel forms of biotechnology achieve the ultimate meshing of the natural, the technical, and the cultural, 'nature stays in place – and in familiar guise' (Daston and Vidal, 2004: 12). Nature, they suggest 'may be a quick-change artist, but it is a hard-working one' (Daston and Vidal, 2004: 14), standing, above all, for order, for the self-evident, for what comes naturally, and ultimately for moral authority.

In the 'post-genomic' context, then, the questions posed by Durkheim and Mauss seem to haunt us still, but perhaps their preoccupations with the naturalness of categories (either via cognitive nature or nature 'out there'), and their suggested answer in the form of looking at 'the social' aspects of boundary making, require a little adjustment today. Opportunities to think through precisely what adjustments are needed, and how we might consider questions of nature/culture relationships arise repeatedly in the re-making of categories seen in many different kinds of genetic technologies. Just as the 'new genetics' in the domain of assisted reproduction is seen by Franklin as an occasion to visit the old problem of what kinship is all about, the Barcoding of Life Initiative, and its genomic methods for assigning specimens to species, provides a chance to revisit debates about nature and culture, the natural and the social, as they inhere in the process of ordering and classifying the organisms within the natural world. Following Franklin, Lury and Stacey (2000), Franklin (2003), Franklin and Roberts (2006), Strathern (1992) and Thompson (2005), I suggest that what may now be needed, and what sociological and anthropological inquiry has in fact taken on, is to look at the formation of *continually emerging connections* between nature and culture in the making of new genomic entities and categories.

This chapter is therefore partly about the entanglement of questions of nature and culture with questions about genomic innovation. What the above authors have found is that, yes, it is true that nature is a very hard-working artist: at a mundane and quotidian level, whether things are deemed 'natural', or not, and how far things deviate from 'what is natural' are questions that lie at the heart of concerns about genomic innovations and their social acceptability. But at a more subtle resolution, what these scholars have been concerned to unpick is the 'traffic' that is occurring between 'natural' and 'social/cultural' categories; together with the observation that these denominations (natural/cultural) no longer cohere as wholes in the post-genomic context. They are only ever fragmentary wholes, always made up in part of the other whilst simultaneously erasing recognition of their fragmentary make-up.

Following the above authors, I look in this chapter at processes of naturalization (or making claims to the natural) in barcoding and, like some of the above authors (eg Thompson, 2005), I examine how naturalization and innovation support one another. In the first half of the chapter I shall attempt to trace the 'principles, forces, and relations' that exist between 'natural' and 'social' fragments which can never wholly account for the innovation at hand (Franklin, 2003: 75; Strathern, 1992: 76). A substantial element of these relations are naturalizing *and* de-naturalizing moves, moves which are selective and shifting and which make up a new 'ontological choreography' around new genetic and genomic technologies (Thompson, 2005). This is a choreography, a dance, between the natural and the cultural/social that provokes intense interest when the issue at stake is innovation both in taxonomy itself, and in 'the species' – a unit/concept, which has, for over two hundred years, been accepted as a fundamental biological entity.

In the latter half of the chapter I turn towards Durkheim and Mauss' second preoccupation: to see what categories 'do'. Here I follow Franklin in looking at the kinds of 'effects, entities and embodiments' that are generated through barcoding (Franklin, 2003). Like the partial wholes that Franklin describes, barcoding has ushered in new taxonomic practices and taxonomic objects (the controversial 'gene-species') that are neither recognizably 'natural' 'social', 'technical' or 'cultural'.

BOLI and COI

The emergence of the Barcode of Life Initiative (or 'BOLI') is traced by most taxonomists to the publication of two related journal articles authored by the Canadian molecular taxonomist Paul Hebert and his colleagues in 2003. Hebert *et al.*'s publications 'Biological Identifications through DNA barcodes' and 'Barcoding Animal Life: cytochrome c oxidase subunit 1 divergences among closely related species' confidently announced the feasibility of a universal system, based around the analysis and 'barcoding' of a fragment of the mitochondrial gene found in all animals, Cytochrome oxidase 1, for the identification of animals-at-large across different taxonomic groupings and across the globe.[4] Claims ran high in these two initial articles, announcing 'barcoding' as the answer to many of taxonomy's early 21st century problems. Barcoding as a universal system would, it was suggested, be based around a 'genomic approach' to taxon diagnosis – an approach that would capitalize on the recent advances of research in sequencing, databasing and information technologies, and one argued to be sorely needed at a time when 'traditional' taxonomic expertise globally was in a state of collapse (Hebert *et al.*, 2003a: 313; Godfray, 2002). Not only would this approach provide a 'reliable, cost-effective and accessible solution' to the acknowledged problem of species identification. Its assembly would, it was argued, also generate important new insights into the diversification of life and the rules of molecular evolution (Hebert *et al.*, 2003a: 313). Hebert *et al.*'s initial papers were confident and assertive of the ability of COI to sustain a new global system of taxonomic expertise:

> Although much biological research depends upon species diagnoses, taxonomic expertise is collapsing. We are convinced that the sole prospect for a sustainable identification capability lies in the construction of systems that employ DNA sequences as taxon 'barcodes'. We establish that the mitochondrial gene cytochrome *c* oxidase I (COI) can serve as the core of a global bioidentification system for animals (Hebert *et al.*, 2003a)

Barcoding's and BOLI's main claim to novelty, however, came not through the 'discovery' of COI as an appropriate 'character' by which to designate species (this had been seen a decade or so before (Folmer *et al.*, 1994, Zhang and Hewitt, 1997) but through:

a) the elevation of COI as *the* gene region that could support a *universal methodology* for the identification, classification and inventorying of species;
b) the promotion of a *singular high throughput, automated and globally standardized sequencing and databasing technology* for doing so;
c) the possibility of releasing *taxonomists all over the globe* from the drudgery and inefficiency of routine identification tasks; and,
d) the promise of *democratizing taxonomy* and rendering its products *accessible* to an unprecedented range of users; or as Barcode of Life literature would have it 'to provide instantly accessible species-level information to 'anyone anywhere' on the globe' (Stoekle, 2003: 797; Ellis, Waterton and Wynne, 2009; Ellis, 2008).

In order to achieve these ambitious social, institutional, technical, 'democratic' and epistemic goals, COI had to be constituted as a natural-technical object that could uniquely support such change. COI, in the early 2000s, was acclaimed by Hebert and colleagues as a universally occurring segment of the genes of natural organisms that could act as a miniscule but powerful 'core' around which a global bio-identification system (initially the entire animal kingdom but imagined ultimately to encompass all species on the planet), could be constructed (Hebert *et al.*, 2003a and b). As the iconic 'pivot' of a larger taxonomic movement ready to revolutionise the 'taxasphere' (Jansen, 2004),[5] COI became heavily naturalized in pro-barcoding literature. At the peak of excitement around the idea/fantasy that barcoding could resolve taxonomic problems on a global scale, those closely involved in barcoding felt strongly that they had found the key that would unlock the secrets of the diversity of life. Genetic barcodes themselves came to be seen as 'embedded in every cell' (Hebert *et al.*, 2003a), and some taxonomists began to acknowledge the indispensability of barcoding as a tool for their craft. In the quotation below, one can see, combined with a (highly modernist) vision of the creation of a global common registry, an unquestioned normalisation of 'the barcode' purportedly allowing scientists to harness a comprehensive indexing system that is 'already established' in nature:

> And so one of the most compelling reasons for barcoding in my mind is that if all of the various tissue collections barcoded their holdings we would suddenly unite this diffuse array of resources under a common registry of genetic sequence accessions. . . . So then the next obvious evolution in this thinking of uniting the holdings of these disparate repositories is sort of make a meta search engine. And so a barcode, that to me, is a real sort of pivotal role, the barcode is allowing us *to harness that infrastructure of information indexing that's already established* . . .[6]

To add to the above example of the naturalization of COI, an insight into the extent to which this quickly occurred within parts of the taxonomic community after 2002, was witnessed by myself and my colleague, Rebecca Ellis, as we sat surrounded by boxes of wasps and ants from far-away places in the world in a typical entomologist's workspace. Casually glancing at the colourful figure represented below which was vividly displayed on a computer on one of the desks in the room at the time (Figure 1), our interviewee commented: 'Yeah, I can't see the world without this now!'

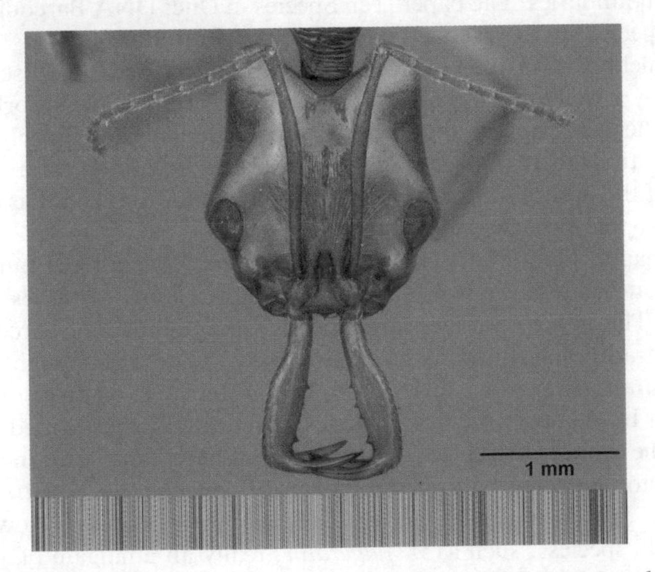

Figure 1: Anochetus goodmani *(a species of ant) and its associated barcode.* April Noble, www.antweb.org

What our interviewee meant by this was that, as an entomologist intimately involved in barcoding, he had come to see and understand the variation between species, and species entities themselves, through barcoding technologies and through barcoding's colourful and intricate representations. Barcoding had become an indispensable, taken-for-granted (or 'naturalized') part of his professional vision (Goodwin, 1994) as an entomologist and a taxonomist.

COI as ordering principle

The naturalizing of COI was of course a process that was absolutely necessary in the early 2000s to support the ambitious social, political, policy, scientific and technical goals associated with barcoding and with BOLI. But the discursive naturalization of COI and the naturalization of barcoding as a bona-fide taxonomic technique also prompted critical questions about this genomic technology – in particular whether species boundaries would, in fact, be rewritten on account of this new genomic method. This was a common concern – and one that was quickly to become a very practical issue. Soon after 2002, COI analysis on the grand scale envisaged by barcoders began to be taken up and promoted by specific naturalists around the world working on different groups of organisms. A frequently cited paper which arose out of a collaboration between practicing field ecologists and the Canadian Centre for DNA Barcoding at Guelph, Canada, served to consolidate the sense that COI analysis could not only assist in quickly, cheaply and accessibly identifying/assigning unknown species to known entities , but could also help to support the accurate rewriting of species boundaries. The paper 'Ten Species in One: DNA Barcoding Reveals Cryptic Species in the Neo-Tropical Skipper Butterfly *Astrapes fulgerator*', based on fieldwork in the Guanacaste Conservation Area in Costa Rica and authored by a well known ecologist, who had devoted his life's work to understanding the Lepidoptera (butterflies) of the Guanacaste Biosphere reserve together with Hebert and colleagues (Hebert *et al.*, 2004), is a spectacular example of the naturalization of COI as *the* character that could assign species to their 'proper' biological identity.

In the paper, the authors claim that their study 'has altered our view of a 'species' that has been known to science for more than 2 centuries' (Hebert *et al.*, 2004). The paper describes how, through the addition of barcoding techniques to traditional morphological taxonomy, what was thought to be one species, *Astrapes fulgerator*, turns out to be harbouring a complex of ten different species. In the paper the word 'species' is repeatedly represented in inverted commas: the authors thereby claim the power of barcoding (used in addition to other taxonomic methods) to destabilize prior conceptions of *Astrapes* and to reassert the existence of new species. They ask with urgency: 'How often are widespread "species", such as *A. fulgerator*, really an amalgam of specialized, reproductively isolated lineages? The answer to this question is crucial to refining estimates of species diversity in the animal kingdom . . .'.

Through the publication 'Ten Species in One', COI was asserted as the new arbiter of natural categories. What the paper did was to denaturalize (deconstruct) existing categories of species and renaturalize (reconstruct) new ones based upon COI analysis. As we shall see later, in this paper and in other research efforts, Hebert *et al.* confirm their commitment to the species as an appropriate taxonomic unit: 'Should the 10 species of *A. fulgerator* identified in this study be formally described despite their morphological similarity? Yes.' (Hebert *et al.*, 2004: 5). Barcoding does not do away with the species unit, it simply (and somewhat controversially, as we shall see later on) shifts its boundaries through the analysis of COI as a supporting (but also defining) 'character'.

From species units to democracy

The controversy that attended the early days of barcoding, however, could not be contained within debates about the accurate designation of species boundaries. Observable in the literature is a constant slippage from a focus on the truth and accuracy of COI analysis to alterative ideas about its (social/policy) efficacy and usefulness. When Prendini, for example, expressed the concern that species designation could not be made on the basis of just one taxonomic tool, COI analysis (Prendini, 2005), Hebert and Barret (2005) disagreed. But in their riposte, they broadened out the whole issue: as well as replying directly about the adequacy of COI as sole indicator, they also included a great number of additional rationales for the support of COI analysis and barcoding. Of primary importance among these was the idea that barcoding would democratize taxonomy:

> By democratizing and speeding access to identifications, barcoding will help to resolve the 'taxonomic impediment' (Janzen, 2004).[7]

Democratization and the speeding up of access to taxonomic information to make it more practical, fit-for-purpose, and usable were constantly repeated benefits of the barcoding method/technology. Indeed an overarching pragmatism underlying barcoding was never far away even when, as above, debates were initiated on technical issues about method. Hebert *et al.* had originated this pragmatic vein of argument in 2003a when they declared that they imagined it possible to create a database for all species on the planet within 20 years for around 1 billion US dollars – a sum considered a snip when compared to the Human Genome Project or other similar big science initiatives. Likewise, they estimated individual specimens could be databased, sequences analysed and entred into barcoding's developing database, BOLD, for less than 2 US dollars per specimen. Barcoders seemed to equate good science with speed, economy, uniformity, standardization, global accessibility and usefulness. Above all, barcoding was heralded as a taxonomy that would be able to identify and document biodiversity faster, cheaper, more efficiently and more *democratically* than ever before. These kinds of rhetorical moves are, of course, what much of genomics is concerned with – an instrumentalizing of DNA for social ends (Franklin, 2003). Indeed, one of the main proponents of Barcoding made it clear to us, in interview, that barcoding was intended to counter the legacy of an esoteric, intellectually-oriented taxonomy:

> I would be very disappointed even if we built these databases and no one used them other than taxonomists, that would be kind of a hollow victory.[8]

As Franklin suggests, admidst this process of instrumentalization, the entities of concern (in our case 'the species') becomes both a social object and a natural phenomenon. Franklin's analysis of the relationship between the social and the natural here is telling for our case. What she suggests is that, even as objects such as the species become at once social and natural (ie both simultaneously), the distinction between social and natural, or biological facts, nevertheless remains crucial, analytically, precisely because it is the enabling condition of instrumentalization. In other words, DNA objects, like the barcode-standing-for-species, become both natural and social phenomenon *through* genomic efforts at instrumentalizing DNA. In this context, efforts to interpret DNA (or COI, or newly designated species) as 'nature' alone and to connect that to other recognized entities (in our case an entity such as 'biodiversity') are an essential part of the innovation, and yet they constantly fall short. In the section above, for example, we have seen that it is impossible for proponents of barcoding to discuss the merits of COI as a sole arbiter of species unit boundaries. COI analysis cannot stand on its own, despite its amazing ability to match unknown specimens to those that are already known. Rather, support for the COI methodology constantly needs to borrow from the language of democratization, utility, accessibility, and so on. Drawing on Strathern, Franklin describes this as a 'mereographic' logic, whereby connecting yet incommensurate orders of phenomena are brought into being. The very incommensurability of DNA with the orders it is supposed to support are part of a logic in which 'DNA will always supersede its social context, in the same way the social context will always supersede DNA'. (Franklin, 2003).

Taxonomic concerns

We shall come back to these theoretical concerns towards the end of the chapter. But first I want to consider a bit more carefully the concerns of other taxonomists (such as Prendini, but there were many others) about barcoding. For practising taxonomists, the issues at stake were two-fold.

First, when COI was naturalized (ie COI was posed as the 'natural' arbiter of species boundaries) critics were quick to point out that designation on this basis could not be valid: that this segment of DNA could not support the (new) delineation of species. Secondly, however, when the argument spilled over into debates about democratization, utility and accessibility (ie COI was posed as ushering in classification that was more invested in technology and politics than nature) barcoding critics argued against this also. Their fear here was that the introduction of barcoding on such a grand and universalizing scale as proposed, paradoxically, would have a destabilizing (rather than a stabilizing, universalizing) effect, and would return taxonomy to the kind of fragmenting 'tower of Babel' condition that had dogged the science's 20th century days. The sense was, if this new genomic method was really about democratization, then this was the wrong way to go about it. It is worth considering some debates in the history of taxonomy that will help explain this two-fold reaction.

From a taxonomic perspective, Hebert *et al.*'s proposals caused a ripple of concern, not to say controversy, within the taxonomic sciences in the early 2000s. But, as Ridley argues, taxonomy is inherently controversial (Ridley, 1986: 2): the announcement of barcoding as a way forward for taxonomy globally only added a further layer to already rabid disagreements within the field. This background of controversy in the everyday landscape of taxonomy is counter-intuitive but nevertheless important for our understanding of the place of nature and that of culture/the social in the daily practices of taxonomists. The normal business of a taxonomist is to arrange plants, animals and micro-organisms into groups according to shared features or 'characters' which themselves are based on some underlying principle of classification (Ridley, 1986). The basic trend is that, 'more and more inclusive groups are defined by more and more generally shared characters, to result in the customary hierarchy of species contained in genera, families, orders, classes, phyla.' (Ridley, 1986: 1). Forms of life are differentiated by their associated characters. Ridley gives some examples: 'horses, for instance, are chordates because at some stage in their lives they possess a hollow dorsal nerve chord, segmented muscles, and a notochord; mammals because they are homeothermic and lactate; ungulates because they are hoofed; and perissodactyls because they support their weight through the centre digit of their feet' (1986: 1–2).[9] But, although seemingly straightforward, all of this is an area of hot dispute. Ridley clearly explains:

> The disagreement of characters is both the fundamental source of all taxonomic controversy and the reason why there is more to classification than the simple definition of groups. Any character will define a group; but different characters do not define the same ones. For instance, the possession of a hollow dorsal nerve chord defines the chordates; but some other character, such as the possession of eyes, defines a quite different and conflicting (and unnamed) group, made up of a miscellany of invertebrates and nearly all vertebrates. So taxonomists are faced with a choice. They have to choose some characters rather than others (1986: 2).

Taxonomists, states Ridley, have had to develop principles (or a philosophy, as he puts it) to underpin character selection, precisely because of the fact that different characters disagree. Taxonomy's main problem at the end of the 20th century, then, was that taxonomy was already riven by two main sets of principles of character selection (the phylogenetic and the phenetic); and these principles had begotten, broadly speaking, three main schools of classification (the evolutionary, the phonetic and the cladistic schools (Ridley, 1986: 163)). Disputes had thrived in the struggle and competition between different taxonomic schools as each school claimed the objectivity of its own philosophy of character selection, all the while facing the same shared problem: 'different principles select different characters: different characters define different classifications.' (Ridley, 1986: 2).

What Hebert *et al.*'s papers did, within this context, was to assert the need for a singular system based on analysis of a single character found in all animals (the gene fragment, COI). Hebert *et al.*'s proposals thus amounted to a wholesale re-thinking of the way in which taxonomy was to work and to provide knowledge about species diversity on the planet. For practising taxonomists, concerns about this new genomic technique were extremely complex. Although competing 'philosophies' of character selection are fiercely defended as unlocking a 'real, unique relationship in nature' (Ridley, 1986: 7), as we have seen, all taxonomists know that different classificatory choices of character (whether the chosen character be a fragment of a gene such as COI, or a morphological trait such as the hollow dorsal nerve chord, or the presence of eyes) define different classifications. Therefore it was difficult for taxonomists to knock down the method based on character selection alone. The problem, as documented in many early critiques of the barcoding proposals, was more concerned with the *singularity* of barcoding's logics, with the idea that this technique, according to a standardized and disciplining process all hinging around the success of COI analysis, was aspiring to be a viable, definitive way of documenting biodiversity world-wide. Taxonomists were concerned both about the all-encompassing, universalizing vision behind this (which looked like it was set to delete diversity amongst the taxonomic sciences – a worrying trend); and about the possible rejection of such a vision within the community which, it was feared , would trigger a return to a situation of chaos, to un-co-ordinated work, and to the famous metaphorical 'Tower of Babel' from which taxonomy 'fit for the 21st century' was struggling to get out (Mallet and Willmott, 2003).

Taxonomists soon began to debate the merits, or otherwise of a 'unitary' web taxonomy (Mallet and Willmott, 2003; Scoble, 2004; Godfray and Knapp, 2004) and some parts of the community vigorously contested the idea that the current crisis in taxonomy might be relieved by DNA barcoding (Mallet and Wilmott, 2003; Seberg *et al.*, 2003; Lipscomb *et al.*, 2003). Whilst many agreed with the need for a way forward into the 21st Century and for an embracing of the digital age, several taxonomists declared concerns about the overtly instrumental tone to barcoding proposals. A problem envisaged by some critics (eg Will *et al.*, 2005; Wheeler, 2008) was that, although the uncertainties of DNA barcoding as a provider of species difference exist within the taxonomic community, these uncertainties might well be sidelined, disappear or seem less accessible and relevant when DNA Barcoding came to be used by others:

> . . . ecologists, behaviourists, conservation biologists, etc., will, without a doubt, move ahead with items identified by DNA barcoding. They will accept the level of non-correspondence of these units to taxa and instead of taxa will use so-called 'gene-

> species' or 'molecular operational taxonomic units' (MOTUs) (Blaxter, 2004), generating a false sense of security that nature has been successfully described (Will *et al.*, 2005: 846).

Will *et al.* go on to say that conservation agencies may use barcoding results in ignorance of the method and momentum of the hype around it and that this could result in 'rash and irreversible mistakes that will impact significant elements of biodiversity' (Will *et al.*, 2005: 848).

Critics of barcoding within the taxonomic community reacted in particular to the singularity of Hebert *et al.*'s vision, and, in many ways, to the naturalization of COI in barcoding rhetoric as the universal 'core' of a new global taxonomic system, The 'stand-off' (Franklin, 2003) between barcoders and critics within taxonomic circles may in fact be typical of new genetic technologies. Critics were not prepared to let a fragment of the gene, COI, stand in for such a powerful vision. And being taxonomists intimately familiar with the nature/culture questions that Durkheim and Mauss posed at the beginning of the last century – they recognized that Hebert's vision was not only a vision of nature, but a vision of culture, too. A 'genetic gap' was thus created, a situation in which DNA (through COI) was understood as both natural and cultural, but could not speak wholly for either, remaining partial and open to ambivalence and dissent.

The necessity and function of partial explanations

If we are interested in the drawing up of boundaries, and in how, and on what basis, this is being done, we do not need to take simply at face value the controversy or stand-off that developed in the early 2000s between barcoders and their critics in the taxonomic community. In the first instance, we have our own controversies to attend to. As Thompson (2005) suggests, it is commonplace, in the post-modern context for scholars to read the fuzziness of the boundaries around nature and culture (and arguments about this fuzziness) as an 'implosion' or 'collapse' of these prior categories: 'All concepts of nature' she suggests, 'including scientific ones, are claimed to be always already shaped, marked, interpenetrated with the imprimatur of culture, and (somewhat less frequently) all concepts of culture are claimed to invoke legitimizing grounds for their systems of classification.' (2005: 174). Thompson also reminds us of the contemporary critics of these kinds of (nature/culture imploding) analyses. These critics, she suggests, undo such implosions, such collapsing of nature/culture boundaries: 'they decry the violation of common sense: there is a real material world that is subject to more or less regular laws, and there is a distinction between truth and falsehood, science and myth' (Thompson, 2005: 174).

Like Franklin, Thompson is highlighting these to-ing and fro-ing debates in the context of new technologies which unsettle established nature/culture boundaries (in both cases, the technologies they are working on happen to be technologies for assisted reproduction in humans). And, also like Franklin, Thompson is looking for a way to describe the partial and temporally unstable ways in which nature and culture *both* do accounting work for these new kinds of technologies. But what Thomson adds to Franklin's description of the mereographic tendencies of these types of new technologies is the insight that the constant accounting work that is done around such new ordering devices (for example, assisted reproduction, or barcoding, for that matter) is a matter of making new meaning, a matter of enacting a much needed ontological dance or choreography within the context of innovation, and a matter of enfolding the downright new into contemporary culture. This constant work is, in the most fundamental sense, a necessary means of renewing culture:

> . . . making distinctions between social and natural roles and facts to ground cultural categories are absolutely fundamental to meaning making in contemporary U.S. society (Thompson, 2005: 176–7).

Not only is such work necessary, it is also incredibly commonplace, mundane, everyday:

> . . . sometimes important political and ontological work is done by denaturalizing what has previously been taken to be natural and deterministic; sometimes the reverse is necessary. (Thompson, 2005: 177)

And lastly, this 'strategic (de)naturalization' is something we can all do, and something we *do* all do: what for some reason seems surprising is that 'ordinary people flexibly manage the choreography between the natural and the cultural' as part of their daily management of the ambivalence of new technologies that affect them (Thompson, 2005: 177).

Partiality and play

Given the discussion above, perhaps it should not surprise us when, in talking to the main proponent of barcoding technology, we find partialities and ambivalences of nature and culture described as inherent within the very kernel of barcoding – the COI gene fragment. As the excerpt from interview below shows,[10] partialities (part-nature/part-culture assemblages) seem to be the fabric from which COI, as a marker deemed fit for the task of successfully designating all animal species to their 'correct' species unit, is made.

> I began to play around with various bits of DNA and looked at a lot of the standard gene targets that people examine – the ribosomal genes, for example. If you look at nuclear genes you soon get reminded of a fact that all evolutionary biologists know – a lot of it evolves really, really slowly and it's not very effective at telling species apart so you've got to look at a lot of DNA to delineate species. By contrast, mitochondrial DNA is really easy to recover and you've got a finite number of targets and, having played around with different genes in the mitochondria in different groups of organisms, I began to develop an understanding of what bits of mitochondrial DNA were most easily liberated. And the other thing that has been a motivation to me from the time that this enterprise began was to try to find a system that anyone

could use – that could be automatable. You know tasks like alignment can consume immense amounts of time, employing secondary structure models trying to align ribosomal DNA . . . I didn't want to do anything complicated, I wanted simple. And when you go for an interior part of a protein-coding gene like COI it turns out that the structure of those proteins are so conserved that it is extremely easy to automate the alignment. There's just no ambiguity in this task.

What this excerpt shows is that the ('natural') speed at which the ribosomal genes evolve was an important consideration in the selection of COI as *the* gene fragment that would signal species difference. But perhaps as, if not more, important, were the other issues that Paul Hebert highlights – the ease of recovery of COI, the need to be able to automate the process so that 'anyone' can do it, the simplicity of the process. Paul Hebert's description of the process of selection of COI as the core of a new system reminds us strongly of the flexible (natural/cultural) accounting that Thompson and Franklin highlight. But what we must also remember in this context of new (reproductive or classificatory) technologies is that the context for those developing the technology is one of innovation – and innovation means intense experimentation, and even play. Paul Hebert, once again, gives us an insight into how much the making of the barcode was a process of imaginative tinkering:

> We [barcoders] make use of every damn scrap of sequence divergence and we then translate it into something that humans can understand, distance. And of course the wonderful thing about the whole distance approach is that it's so damn fast. You can sit in front of a computer screen and you can build a tree in a matter of minutes with a thousand things in it. . . . (Hebert transcript, November, 2006: 13)

Hebert is playing, in a sense, not only with issues of 'distance' versus other ways of representing species divergence, but with the whole idea of what taxonomy is for. He articulates this thus: '. . . really you know the question is one of intellectual purity versus pragmatism'. In the process of tinkering, Hebert is asking: is taxonomy an intellectually pure pursuit or is it a pragmatic science to deliver instrumental gains for the perceived needs of humanity?

What this underlines is that the ambivalences and partialities concerning how we account for this new genomic way of ordering nature are as much to do with the wider post-genomic technoscientific context as they are to do with the technology itself. As many have documented, the genomic sciences are typically those that are self-consciously just as responsive to economic and cultural needs, ideologies and imaginaries as they are to epistemic ones (Sunder, 2006). Within this wider post-genomic context, taxonomic innovation, as Hebert eloquently points out, means that the whole 'question' or philosophy of the discipline of taxonomy is up for grabs.

Barcoding's embodiments and effects

In the early 2000s, barcoders were declaring that the relationships around the practice of taxonomy were changing so significantly that the practice commonly known as taxonomy itself had to shift and expand. Barcoding was reinventing the rules of the game in taxonomy, re-writing its practices not for taxonomists, but for 'humanity', not for discreet phylogenetic aims but for understanding, as fast as possible, the extent of biological diversity on the planet. Barcoding was portrayed as a serious yet playful, maybe even a maverick, genomic science, self-consciously oriented to new instrumental ends. Barcoding was often naturalized – but as a useful science – one that would deliver tangible results fast and efficiently to its purported users.

A central ambiguity within BOLI followed from this. When barcoding was first announced as a possible system for taxonomy as a whole (Hebert *et al.*, 2003 a and b) barcoders suggested that their proposed system would be both a system of identification and provide insight for the classification of organisms (initially the animal kingdom) in an evolutionary perspective:

> When fully developed, a COI identification system will provide a reliable, cost effective and accessible solution to the current problem of species identification. Its assembly will also generate important insight into the diversification of life and the rules of molecular evolution. (Hebert *et al.*, 2003a: 313)

However, barcoders were rapidly criticized for assuming they could build a new phylogeny on just one small fragment of the genetic material of all animals on earth (Moritz and Cicero, 2004; Prendini, 2005). They reacted by reining in their promises, assuring the taxonomic community and others that their main preoccupation was actually not with 'the branches' (the historical evolution of species) but with 'the tips' (observations of character difference) in the biological species hierarchy. What barcoders *could* assert, post 2003, was that a short COI gene sequence taken from a standardized position on the mitochondrial genome of a specimen could be converted through the use of sequencing and databasing technologies into an easily represented, simple, barcode available digitally to everyone. This barcode could be 'blasted' or checked against all DNA barcodes in the BOLI database and the identity of the specimen from which it was derived could be revealed through the detection of difference from other barcodes already banked. This process involves deriving species identification from DNA, rather than from morphological characters. However, unlike DNA taxonomy which is prepared to create a new taxonomy based around genetic information alone, the sequences derived from barcoded specimens are usually fitted back to the Linnaean binomial classification system retrospectively (Vogler and Monahan, 2006). As a result, DNA, in barcoding, does not constitute a new 'scaffold' for the grouping of living organisms; it is simply a route through which researchers can gain quicker access to established Linnaean binomial species groupings.

For this reason, barcoding was noted to be a 'conciliatory approach' (Vogler and Monahan, 2006), using valuable sequence information only to aid the study of curated voucher specimens and to align these with previously established Linnaean names. As le noted elsewhere (Waterton, Ellis and Wynne, forthcoming) the conciliatory approach was seen by critics to bypass all the intellectual excitement associated with taxonomy – including the possibility that DNA-based information could itself be used as a powerful tool for testing species concepts, for questioning the biological nature of species, for understanding the evolution of species and for predicting species divergence (Wheeler, 2008; Will *et al.*, 2005; Vogler and Monahan, 2006). Although barcoding was heralded both initially and still, after 2003, as a way of revolutionizing taxonomy, many considered the approach nothing more than a 'by-product' of the Linnaean classification system, rather than a taxonomic system itself (Vogler and Monahan, 2006). As such, barcoding established new and controversial practices, and also, its critics argued, created distracting new objects. In a damning critique of barcoding proposals, Will *et al.*, 2005 described barcoding as a blind and pre-emptive harvesting of sequences, involving the dangerous establishment of a new hybrid – the compromising 'gene-species'. This was an object that blended, in confusing ways, the open-ended possibilities associated with DNA taxonomy with the much more conservative epistemic task of matching specimens to traditional Linnaean species nomenclature.

Barcoding itself therefore did not fit into the existing classification of accepted taxonomic practices; preferring to focus on issues of standardization, accessibility and democratization of taxonomy rather than issues of phylogeny, hypothesis testing and possible new DNA-based orderings. But what is interesting for our purposes in thinking through the emerging connections between nature and culture in the making of 'new' genomic entities and categories is the way that barcoding both naturalized COI as a universal marker for the identification of all species on the planet and, paradoxically, had the effect of temporarily denaturalizing the whole global enterprise of collecting, analysing and classifying natural entities.

Conclusion

Maciel describes the present time as a 'hypertextual' era – an era in which 'the speed and multiplicity of information explicitly de-authorizes and disestablishes the very idea of classification, demanding a re-shaping of knowledge beginning from a more open, dialogical and even paradoxical perspective . . .' (2006: 49). Barcoding, it would seem, is a child of this era. Barcoding's short history could not have evolved in the way that it has without the existence of an acknowledged 'crisis' in taxonomy, a discipline which, in the early 2000s, was facing both internal and external critique of its inadequacy as a science that might in some way alleviate the accelerating loss of global biodiversity (House of Lords, 1992, 2002; The Royal Society, 2003). At the same time our very familiarity with the term 'barcoding', most frequently understood as a technology used for coding commercial products (with all its associated connotations of audit, tracing, stocktaking and tracking), subtly highlights the kind of borrowing, cross-overs and translations that are underway as the bulky, fleshy world of biology and specimens meets the virtual and digital media of code and information, via genomics.

I have argued in this chapter that debates around barcoding both naturalize COI and destabilize the natural basis of barcoding. In attempting to harness the power of genomics towards instrumental ends, barcoding encompasses several seeming contradictions. We have seen that barcoding aims to revitalize taxonomy through an unprecedented universalism, yet the search for the universal marker is a limited and pragmatic (or 'conciliatory') one. We have also seen that barcoding is simultaneously both a very conservative taxonomic move (using DNA methods but reverting to a Linnaean system of ordering) *and* a radical break with taxonomic convention through its orientation to policy concerns about the global biodiversity crisis. These apparent contradictions, I suggest, following Franklin *et al.* (2000) and Franklin (2003), also harbour typical processes of naturalization, de-naturalization, and re-naturalization as witnessed in many areas where genomics and biology converge (Franklin *et al.*, 2000: 9–10; Franklin, 2003). What Franklin suggests is that the biological, in these kinds of contexts where DNA is being harnessed for instrumental ends, is being asked to do a very particular kind of 'merographic work' (Strathern, 1992) whereupon it becomes a category in which nature is both unavoidably present, and yet increasingly absent. The biological in this sense connects ideas of nature and society and is composed of distinct orders that belong to different, distinct, and irreconcilable wholes. Increasingly, the biological refers to 'a combination and division of phenomena' that not only requires new metaphors, *but also comes to embody them* (Franklin, 2003: 691).

Thinking back to the preoccupations of Durkheim and Mauss – the question of how, and on what natural/cultural basis, we make classes out of things – we continue, however, to be gripped by their insight that 'neither the tangible world nor our mind gives us the model' (Durkheim and Mauss, 1970 [1903]: 8). It seems important, however, to continue to question how new orderings, including orderings of the natural, are being accounted for. As I suggested earlier, the making of new categories (and the subsequent disturbance of old ones) means that the question is not just about the natural and the cultural as they inhere in practices of ordering. It is partly about how we 'do' nature and culture around innovation, in the making of the new. Thompson describes this perfectly within the context of new reproductive technologies:

> Given . . . that technological innovation and cultural history implicate each other so strongly, it is no wonder that progressive cultural critics cannot decide whether the new reproductive technologies are best judged as innovative ways of breaking free of bondage to old cultural categories of affiliation or whether they are best denounced as part of a hegemonic reification of the same old stultifying ways of classifying and valuing human beings. The technologies are fundamentally both. (Thompson, 2005: 177–8).

Just as the technologies are fundamentally both, so are their effects and embodiments. We have seen in this chapter that new embodiments, and new entities do arise out of the processes of naturalization, denaturalization and renaturalization we are witnessing. For barcoders, the kinds of embodiments and effects that are desired are databases full of barcoded specimens, being accessed globally by crowds of imagined new users, and being used in the service of a kind of fantasy global biodiversity protection. For barcoding critics, such entities and effects include the compromising gene-species and the possibility of nature being poorly described and categorized, albeit in the name of a technologically sophisticated democracy. These embodiments and effects, like the processes that give rise to them are complex, neither wholly natural nor wholly cultural: we will need to get used to these new kinds of entities, and get used to new ways of accounting for them, in the post-genomic era.

Notes

1 On the 'classificatory function' Durkheim and Mauss assert 'Logicians and even psychologists commonly regard the procedure which consist in classifying things, events and facts about the world into kinds and species, subsuming them one under the other, and determining their relations of inclusion or exclusion, as being simple, innate, or at least instituted by the powers of the individual alone. Logicians consider the hierarchy of concepts as given in things and as directly expressible by the infinite chain of syllogisms. Psychologists think that the simple play of the association of ideas, and of the laws of contiguity and similarity between mental states, suffice to explain the binding together of images, their organization into concepts, and into concepts classed in relation to each other'.

2 Needham, for example, criticized heavily the social causality implied in Durkheim and Mauss's arguments about classifications which he saw as overly deterministic (Needham, R. 1970 [1963], 'Introduction'. In Durkheim and Mauss, 1970 (1903) *On some Primitive forms of classification: contribution to the study of collective representations*'. London: Routledge (1970), pp. vii–xiviii. And we can go much further back in history to the ancient Greeks and the work of Plato to see documented the recognition that classifications are not all about 'the essential identity of the thing' or 'equivalence'; they are also about 'association', or as Durkheim and Mauss would put it 'relations'. (Boyne, 2006, Classification Theory, *Theory Culture and Society*, 23 (2–3): 21–50).

3 The questions posed by Durkheim and Mauss have been followed through in anthropology, history of science and sociology, see, for example a special issue in *Theory Culture and Society* that makes frequent reference to this foundational text: Boyne, 2006, 'Classification Theory', *Culture and Society*, 23 (2–3): 21–50; also Atran, S., 1990, *Cognitive Foundations of Natural History: Towards an Anthropology of Science,* Cambridge: Cambridge University Press. In anthropology, questions about classification have been especially re-worked in the area of ethnobotany: see, eg, Sanga and Ortelli, 2004, especially chapters by Maddalon, Berlin, Ellen, Longo, Trumper and Iannaccaro; Sanga, G. and G. Ortelli, (eds), *Nature Knowledge: Ethnoscience, Cognition and Utility*, New York and Oxford: Berghahn Books.

4 It was not known at the time whether the methodology proposed would 'work' as well for the plant kingdom. Initial claims for a universal method were therefore restricted to animals.

5 A handy term coined by Daniel Jansen for denoting the diverse people materials, methods, collections, involved in doing taxonomy (Jansen, 2004).

6 Interview with Investigator at the Canadian Centre for DNA Barcoding and Chair of the CBOL Database Working Group, 2007. Italics added.

7 Reply to the comment by L. Prendini on 'Identifying spiders through DNA barcodes', Paul D.N. Hebert and Rowan D.H. Barrett, *Can. J. Zool.* 83: 505–506 (2005).

8 Interview with CBOL Representative, Washington 2007: 13.

9 For the most comprehensive overview see Hull, David 1988 Science as a Process: An Evolutionary Account of the Social and Conceptual Development of Science. University of Chicago Press.

10 The excerpt is taken from an interview carried out between my colleague Rebecca Ellis and one of the original 'authors' of barcoding, Dr Paul Hebert. Hebert Transcript, November 2006: 13.

References

Atran, S., (1990), *Cognitive Foundations of Natural History: Towards an Anthropology of Science*, Cambridge: Cambridge University Press.

Bowker, G. and S.L. Starr, (1999), *Sorting Things Out: Classification and its Consequences*, Cambridge, Massachusetts: The MIT Press.

Boyne, R., (2006), 'Classification', *Theory, Culture and Society*, 23(2–3): 21–50.

Clarke, A.E. and J. Fujimura, (eds), (1992), *The Right Tools for the Job: At Work in Twentieth Century Life Sciences*, Princeton: Princeton University Press.

Daston, L. and F. Vidal, (2004), (eds), *The Moral Authority of Nature*, Chicago, IL: University of Chicago Press.

Durkheim, E. and M. Mauss, (1970 [1903]) *On some Primitive forms of classification: contribution to the study of collective representations*, London: Routledge.

Ellis, R., (2008), 'Rethinking the value of biological specimens: laboratories, museums and the barcoding of life initiative', *Museum and Society*, 6: 172–191.

Ellis, R., C. Waterton and B. Wynne, (2009), 'Taxonomy, Biodiversity and their Publics in 21st Century Barcoding', *Public Understanding of Science*, OnlineFirst, published on July 3, 2009 as doi:10.1177/0963662509335413.

Folmer, O., M. Black, W. Hoeh, R. Lutz and R. Vrijenhoek, (1994), 'DNA primers for amplification of mitochondrial cytochrome *c* oxidase subunit I from diverse metazoan invertebrates', *Mol. Mar. Biol. Biotechnol.*, 3: 294–299.

Franklin, S., (2003), 'Re-thinking Nature-Culture: anthropology and the new genetics', *Anthropological Theory*, Vol 3(1): 65–85.

Franklin, S., C.E. Lury and J. Stacey, (2000), *Global Nature, Global Culture. Gender, Theory and Culture*, London: Sage.

Franklin, S. and C. Roberts, (2006), *Born and Made: an ethnography of preimplantation genetic diagnosis*, Princeton: Princeton University Press.

Goodwin, C., (1994), 'Professional Vision', *American Anthropologist*, 96(3): 606–633.

Godfray, H.C.J., (2002), 'Challenges for taxonomy', *Nature*, 417: 17–19.

Godfray, H.C.J. and S. Knapp, (eds), (2004), 'Taxonomy for the 21st Century: Special Issue', *Philosophical Transactions of the Royal Society, series B, Biological Sciences*, 359: 559–739.

Hebert, P.D.N., A. Cywinska, S.L. Ball and J.R. deWaard, (2003a), 'Biological identifications through DNA barcodes', *Proc. R. Soc. Lond.* B 270: 313–321.

Hebert, P.D.N., S. Ratsingham and J.R. deWaard, (2003b), 'Barcoding animal life: cytochrome *c* oxidase subunit 1 divergences among closely related species', *Proc. R. Soc. Lond.*, B 270(Suppl.): S96 S99.

Hebert, P., E. Penton, J. Burns, D. Janzen and W. Hallwachs, (2004), 'Ten species in one: DNA barcoding reveals cryptic species in the neotropical skipper butterfly *Astraptes fulgerator*', *Proc Natl Acad Sci USA*, 101: 14812–14817.

Hebert P.D.N. and R.D.H. Barrett, (2005), 'Reply to the comment by L. Prendini on "Identifying spiders through DNA barcodes"', *Can. J. Zool.*, 83: 505–506.

House of Lords, Select Committee on Science and Technology, (1992), *Systematic Biology Research 1st report*, HL Paper 22-1. London, UK: House of Lords.

House of Lords Select Committee on Science and Technology, (2002), *What on Earth? The threat to the science underpinning conservation*, 3rd Report, HL Paper 118 (i), London, UK.

Hull, David, (1988), *Science as a Process: An Evolutionary Account of the Social and Conceptual Development of Science*, Chicago, IL: University of Chicago Press.

Jansen, D., (2004), 'Now is the Time', *Phil. Trans. R. Soc. Lond*, B., (2004) 359: 731–732.

Lipscomb, D., N. Platnick and Q. Wheeler, (2003), 'The Intellectual Content of Taxonomy: a comment on DNA taxonomy', *Trends in Ecology and Evolution*, 18, 2: 65–66.

Maciel, M.E., (2006), 'The Unclassifiable', *Theory Culture and Society* (Special issue on Problematizing Global Knowledge-Classification/The Collection) 23(2–3): 47–50.

Mallet, J. and K. Willmott, (2003), 'Taxonomy: renaissance or Tower of Babel?' *Trends in Ecology and Evolution*, 18: 57–59.

Moritz, C. and C. Cicero, (2004), 'DNA barcoding: promise and pitfalls', *PLoS Biol*, 2: 1529–1531.

Needham, R., (1970 [1963]), 'Introduction', in Durkheim E. and M. Mauss, (1970 [1903]) *On some Primitive forms of classification: contribution to the study of collective representations*'. London: Routledge: vii–xiviii.

Prendini, L., (2005). 'Comment on "Identifying spiders through DNA barcodes"', *Can. J. Zool*, 83(3): 498–504.

Rabinow, P., (1996), *Making PCR, A Story of Biotechnology*, Chicago, IL: University of Chicago Press.

Ridley, M., (1986), *Evolution and Classification: the Reformation of Cladism*, New York: Longman.

(The) Royal Society, *Measuring Biodiversity for Conservation*, Policy document 11/03, ISBN 0 85403593 1; www.royalsoc.ac.uk; London, August 2003.

Rose, N., (2007), The Politics of Life Itself: biomedicine, power, and subjectivity in the twenty-first century, Princeton: Princeton University Press.

Rose, N. and C. Novas, (2004), 'Biological citizenship', in Ong, A. and S.J. Collier (eds), *Global assemblages: technology, politics, and ethics as anthropological problems*, Oxford: Blackwell Publishing: 439–463.

Sanga, G. and G. Ortelli, (eds), *Nature Knowledge: Ethnoscience, Cognition and Utility*, New York and Oxford: Berghahn Books.

Scoble, M.J., (2004), 'Unitary or unified taxonomy?' *Philosophical Transactions of the Royal Society of London B,* 359: 699–710.

Seberg, O., C.J. Humphries, S. Knapp, D.W. Stevenson, G. Petersen, N. Scharff and N.M. Andersen, (2003), 'Shortcuts in systematics? A commentary on DNA-based taxonomy', *Trends in Ecology and Evolution*, Vol. 18(2): 63–65.

Stoeckle, M., (2003), 'Taxonomy, DNA, and the bar code of life', *BioScience*, 53: 796–797.

Strathern, M., (1992), *After Nature.* Cambridge: Cambridge University Press.

Sunder Rajan, K., (2006), *Biocapital: The Constitution of Postgenomic Life*, Durham, NC: Duke University Press.

Thompson, C., (2005), *Making Parents: The Ontological Choreography of Reproductive Technologies*, Cambridge, Mass: MIT Press.

Vogler A. P. and J. Monahan, (2006) 'Recent advances in DNA taxonomy'. *J. Zool Syst Evol*, 29 July 2006: 1–10.

Waterton, C., R. Ellis and B. Wynne, (forthcoming), *Barcoding Nature: shifting taxonomic practices in an age of biodiversity loss*, London: Routledge.

Wheeler, Q.D., (2005), 'Losing the plot: DNA, barcodes, and taxonomy', *Cladistics*, 21: 405–407.

Will, K.W., B.D. Mishler and Q.D. Wheeler, (2005), 'The perils of DNA barcoding and the need for integrative taxonomy', *Syst Biol*, 54: 844–851.

Wheeler, Q.D., (ed.), (2008), *The New Taxonomy*, The Systematics Association Special Volume Series 76, New York: CRC Press.

Zhang, D-X. and G.M. Hewitt, (1997), 'Assessment of the universality and utility of a set of conserved mitochondrial primers in insects', *Insect Mol. Biol*, 6: 143–150.

Part Six
Theorizing Nature Through Genomics

Genomic natures read through posthumanisms

Richard Twine

This chapter sets out arguments and histories of posthumanisms to use their tensions as a way to think about 'nature' in genomics. I argue that genomics frames 'nature' ambivalently in ways that are both faithful to *and* undermining of Enlightenment understandings of nature. New biotechnological innovations and their associated imaginaries have become the scene for much speculation on the ontological status of the 'human'. Genomics is paradoxical for the human (and humanism) since on the one hand the human genome project professed to reveal the 'human',[1] but, simultaneously, prior assumptions about a notion of human nature appear increasingly fragile in the face of genomic visions of human 'enhancement'. Yet it would be a considerable mistake to assume that these material incursions upon both 'nature' and bodily flesh have been the prime reason why the 'human' has been called into question. I contend that this point signals a distinction within the posthumanist terrain – which is both historical and political – between transhumanism and critical posthumanism. Whilst neither of these terms, as we shall see, encompass a wholly coherent set of ideas and certainly feature differences of note within their terrain, the intention here is to examine their distinction as a useful way to think how 'nature' is conceptualized in genomics. We should be careful from the outset when thinking about 'genomics and nature' not to settle into a notion of 'nature' as somehow separate from 'culture' or the 'human'. I use the term genomics broadly not merely to refer to sequencing projects but also to include related fields such as comparative genomics, metagenomics and hybrid embryos.

The various constellations of posthumanist thought are invested in rethinking the 'human' albeit in different ways. I shall argue that transhumanism is generally faithful to Enlightenment understandings of 'nature' in dualist terms, whilst the critical posthuman project generally attempts to unravel dualistic ontology.[2] The term 'humanism' itself has various meanings during different historical periods but is characterized by moral ambivalence as both a supposed foundation for human freedom from oppression and superstition as well as a partial and exclusionary discourse. My specific deployment here, common to the overlapping fields of critical posthumanism and animal studies, is of humanism as a reduction of value and agency to the 'human', a curiously centred and bounded category that has elevated itself by contrast to the 'animal' and drawn upon ideas of animality to essentialize human difference. This partial critique of humanism is sympathetic to a view of humanity as

> neither an essence nor an end, but a continuous and precarious process of becoming human, a process that entails the inescapable recognition that our humanity is on loan from others, to precisely the extent that we acknowledge it in them. For those 'westerners' whose humanness is mortgaged to the suffering and labour of an innumerable 'Other', the recognition cannot be comfortable or merely reflective (Davies, 1997: 132).

Yet the ending to this quote ironically speaks to a humanism that excludes nonhuman suffering and labour. It is precisely this exclusion, as we shall see, that certain versions of a critical posthumanism address themselves to.

Transhumanism may be briefly defined as the belief that the human race should be 'enhanced' using technological means.[3] This discourse, which sometimes appears like a caricature of humanism with its commitment to the technological mastery[4] of 'nature' and celebration of the autonomous subject, varies in the strength of its argument as well as its discursive location. Transhumanist ideas (not necessarily self-proclaimed as such) are to be found in some science fiction and some bioethics for example. The terrain is further confused by the propensity for transhumanism to be labelled or self-identified as posthumanism.

However distinctly *critical* posthumanisms are also diverse and can place a different emphasis on the extent to which they critique human power relations vis-à-vis the nonhuman. Thus I devote space below to considering whether critical posthumanism is also a politicization of the nonhuman. In contrast to transhumanism, critical posthumanisms are better seen as

> a political-analytical perspective than a historical moment and as extending the well-established critique of the reductive but effective categories of human/animal, nature/culture (Castree and Nash, 2006: 502).

Overall I take the methodological approach of *including* transhumanism, critical posthumanisms and also antihumanism as all residing within the posthuman, speaking to a plural and partially overlapping terrain of posthumanisms.

The first section illustrates that the posthuman pre-dates the discourse of transhumanism and so is an argument against the singular colonization of the posthumanist field by transhumanism. The second attempts to clarify the uses of posthumanism through reference to recent literature, and contributes to a specifically non-anthropocentric[5] critical posthumanism. I am mindful of the non-innocent practice of typology but aim to achieve some conceptual clarification of a complex field as well as to underline the critical posthumanist critique of transhumanism. The third section explores the extent to which genomics and related biosciences reinforce various strands of posthumanism. The cases of hybrid embryos and comparative farm animal genomic databases are employed in order to explore this question. These examples are chosen for the way in which they bring the human and nonhuman into close proximity and for their illustration of how contemporary knowledge production takes place through a questioning of species boundaries. This process is examined for its potential implications and how these may or may not chime with varying discourses of posthumanism. First I introduce posthumanism via a consideration of human decentring.

Human wounds

> We're undone by each other. And if we're not, we're missing something (Butler, 2004: 19).

Part of the difference between posthumanisms as a whole is that more critical strands have been constituted by heterogeneous themes of 'human' *decentring* which begin as far back (at least) as the ramifications of the heliocentric cosmology of Nicolaus Copernicus in the early part of the 16th century. Badmington (2000: 5–6) provides a useful account of some of these decentring moves from quite different scholars. To the decentring of the Earth from the centre of the universe he adds Marxism, Darwinism and psychoanalysis as having provided an important contribution to posthumanist thought. Thus for Badmington the Marxist critique of Hegelian idealism, positing instead the social and material construction of consciousness, psychoanalysis that questioned both the autonomy of the self and the rationality of 'man', and the 'shocking blasphemy' of Darwinism that would return the 'human' *alongside* other animals, have all been important sources of posthumanism. Derrida names these woundings; the Copernican, the Freudian, and the Darwinian as the three traumas of humanity (in Wolfe, (ed.), 2003: 138–9). Haraway adds a fourth – the realization that various nonhuman actors are lively (Gane, 2006: 141) or agential. These 'traumas' and their potential for inspiring conceptual shifts are neither definitive nor final. They have different emphases; they may undermine a belief in human omniscience, and/or do work to erode a sense of self as bounded or atomistic. Butler names grief and desire as two of the more obvious embodied emotional relations which call into question accounts of the self as always autonomous or in control (2004: 19). Both environmental and animal ethics discourse have also destabilised the 'centred human' and are accompanied by more recent developments within various fields of science.

Animal cognition research throws up regular, seemingly unexpected, nonhuman abilities such as the magpie becoming the first non mammal to be able to recognize 'itself' in a reflection.[6] Or, in an example that has specifically enrolled genomics; interdisciplinary work around human evolution constructs *Homo sapiens* as just one of several *Homo* species. Both *Homo neanderthalensis* and *Homo floresiensis*[7] were contemporaneous with *Homo sapiens*, dying out approximately 24,000 and 12,000 years ago respectively. The Neanderthal Genome

Project seeks to clarify the relationship between *Homo sapiens* and *Homo neanderthalensis*[8] with a special interest in whether an admixture occurred between the two species. Genomics has thus become important to the debate within human evolution as to whether *Homo sapiens* originated solely from a relatively small population in what is now Africa or whether hybridization took place between different sub-species of *Homo sapiens* across different geographical areas. For example, Eswaran *et al.* (2005) argue that genomic analysis suggests evidence for the assimilation of 'archaic' humans (see also Garrigan and Hammer, 2006) into the modern human genome (even if it is arguably problematic to conceive of such a homogenous 'modern human genome'). Genomics has also been drawn upon to argue for a near catastrophic human population decline in the late Pleistocene period. Genomic research around the 'pre-human' has the capacity to unfold significant posthumanist decentrings. We see the possibility of a rather contingent, contested, multiple and arbitrary 'human'; one that emerged out of other species and sub-species and one whose genetic diversity has been impacted upon at various points of socio-natural population pressure. Whilst genomics inspired human evolutionary research shares with the Darwinian rupture a placing of the 'human' alongside other species it also acts to further decentre an understanding of a 'human nature'; now better understood as something both pointedly historically situated and unstable.

The application of genomic knowledge to the study of human evolution fits with the broader rationale of species genome sequencing efforts as attempts to clarify boundaries and relations between the 'human' – constructed as *Homo sapiens* – and other animal species. Beyond the knowledge of human development and self-understanding implicit here is the search for the 'something special' of the 'human' which could situate genomics as a science interested in buttressing a sense of human exceptionalism. But as we shall see the picture is somewhat more complex.

Mapping the posthuman

a) Antihumanist influences on critical posthumanism

Although the use of genomics in human evolutionary theory stands as a relatively recent example of human decentring, others, as noted above, are ironically at least as old as the emergence of Enlightenment humanism. Within these various decentrings we can begin to glimpse some of the diversity within posthumanism. Firstly it is misleading to posit posthumanism simplistically as a radical break from humanism. Whilst several of the aforementioned intellectual trends have served to decentre the human it is also the case that they have been important to various forms of humanist thought. Davies points out a way to start thinking about this contradiction:

> On one side humanism is saluted as the philosophical champion of human freedom and dignity, standing alone and often outnumbered against the battalions of igno-

> rance, tyranny and superstition. . . . On the other it has been denounced as an ideological smokescreen for the oppressive mystifications of modern society and culture, the marginalisation and oppression of the multitudes of human beings in whose name it pretends to speak . . . (1997: 5).

From Davies' quote we can see how Marxism, Darwinism and psychoanalysis have all been a part of this humanist aspiration for 'human freedom'. But several of these decentring moves have also reinscribed other more centralizing, partial visions; leading to critiques, for example, of the gendered character of psychoanalysis. These trends which have been critiqued and re-worked in the context of post-structuralist thought have had methodological and theoretical import for the social sciences and humanities. Thus for some these could be more properly attached to the notion of antihumanism rather than speak for the whole posthumanism terrain. Here I understand antihumanism as ontological projects against human essence and the notion of the autonomous human subject.[9] This is an important part of posthumanism generally, but it cannot be said to encompass all expressions of the posthuman, being something of an anathema to transhumanism, but important to many critical posthumanisms.[10]

The point here is that antecedents of such critical posthumanisms are to be found in the histories of humanism. The exclusionary practices of humanism, which Davies alludes to above, have been possible, in a conceptual sense, due to the unsatisfactory resolution of reason/nature dualism. Arguably *post*humanism only comes into its own when it labours to reformulate, critically, the hierarchy of this dualism, to intervene against the productivity of 'nature' and 'animality' as discursive pathways to dehumanization, whilst *simultaneously* questioning the original implications of a 'dehumanization' discourse. This is the difference within the posthuman which Braidotti alludes to as a

> twofold dimension within this complex category: the first concerns the philosophical post-humanism of the post-structuralist generation; the second is a more targeted form of post-anthropocentrism that is not as widespread (2006: 197).

We must then augment Badmington's (2000) narrative of critical posthumanism with further interlinked influences, for example feminism and environmentalism, strands of thought that are significant to Braidotti's second dimension of critical posthumanism. Feminism continues to call to account the colonization of the 'human' by hegemonic understandings of the masculine, an effort which had a substantial reflective consequence for noting the shortcomings of a liberal feminism that failed to adequately answer the question 'equality with what?'.[11] I frame the challenges to the 'human' represented by feminism[12] and environmentalism as interlinked here because the masculine colonization has decisively also involved a particular externalised relation to nature. I take the meaning of anthropocentrism from Plumwood[13] (1996) as a human centrism that denies interests to other species, and as interlinked with other forms of centric thought most notably androcentrism and ethnocentrism. Environmental philosophy and eco-social theory extend prior decentrings of the 'human' in a quite different sense to the Marxist and Freudian strands noted above. These have more in common with the decentrings of both Darwinism and romanticism in that the human is conceptually (re)located as embedded in both bodies and ecologies. Again there is a methodological point to be made here. For the Enlightenment explication of the 'human' generally promoted mediated dualistic understandings of both the 'social' and of the medical apprehension of human disease and flourishing. Moreover there is a deep ethical point to be made about human exceptionalism, of human supremacism and separation. Rawles refers to this overly instrumentalised ethical construction of 'nature' as 'a pre-Copernican view of ethics; the values equivalent of believing the sun spins round earth' (2007: 14). So for what we could frame as an overtly ecologically inflected critical posthumanism[14] (and implicitly for feminist posthumanism) there is a further challenge to what Ehrenfeld referred to as the 'arrogance of humanism' (1981) that takes issue with the view of the 'human' as the measure of all value, the axiomatic pinnacle of evolution, and the taken for granted mastery of other species. Grasping the discursive form of reason/nature dualism and its repetitive mappings onto related dualisms illustrates adeptly the partiality of the Enlightenment 'human' historically defined in opposition to domains constructed as 'closer to nature'.

As one may note, there is potentially an interesting conflict between this ecologically inflected critical posthumanism and antihumanism. For although the former is not necessary making very strong claims for a 'human nature', by underlining human materiality and our interconnections with bodies and ecologies, it could be seen in one sense to be counter antihumanist (arguing for a universal embedded context for the human) and in another sense confirmatory to antihumanism (since our dependency upon bodies and ecologies is another blow to the notion of individual autonomy).

Although this critical posthumanism makes an epistemological, ontological and ethical break it once again should not necessarily be seen as radically opposed to all variations of humanism. Indeed specific expressions of critical posthumanism such as certain ecofeminisms (eg Plumwood, 1993) posit an integrated theory of power connecting nature mastery to that of class, 'race' and gender; thus chiming (though superseding) partly with humanist emancipatory foci, and certainly not abandoning a model of 'progress'. Such examples undermine the temptation to equate all accounts of posthumanism with misanthropy; *instead* perceiving a human benefit in its own decentring in the same way as feminist critique may try to argue for the experiential benefit to men of learning feminist critique.

How does this inflection of posthumanism gel with other typologies of the posthuman terrain? Self-labelled versions of critical posthumanism (eg Badmington, 2003; Thacker, 2003; Didur, 2003; Rossini, 2006) that partly resemble the aforementioned ecologically inflected posthumanism have served as modes of distancing from what may be seen as naively techno-celebratory accounts of the posthuman. This is Thacker's two-way distinction between what he terms on the one hand as 'extropianism' (more commonly referred to as transhumanism) and on the other a wide range of critical posthumanisms which 'offer a more rigorous, politically and socially rooted body of work . . .' (2003: 80). Writing specifically on genetic technologies Didur aligns her critical posthumanism with that of Hayles (1999) and Haraway (1997),distancing it from the corporate interests of Monsanto as well as the transhumanist arguments of German philosopher Peter Sloterdijk:[15]

> While Sloterdijk and Monsanto imagine genetic engineering as humanity's way of perfecting nature and thus undermining the originary and hierarchical divide between nature and culture, Man and machine, critical posthumanism questions the view that there was ever an originary divide between these things in the first place (2003: 101–2).

There seems to be a concern that such transhumanist discourse has begun to colonize the posthuman terrain. Indeed Donna Haraway, even though she acknowledges the diversity of posthuman writing and is often cited as a pivotal part of it, has stopped using the word posthuman precisely over such fears of transhumanist appropriation, preferring instead to focus upon a discourse of species (Gane, 2006: 140).[16] Hayles, in contrast, is more optimistic for the posthuman. Her posthumanism is also argued in difference to transhumanism but she is diligent to make clear that there is a promising posthuman terrain outside of transhumanist concerns (1999: 283).

b) Considering transhumanism

What of transhumanism? Why does it inspire such objections? These as we shall see take different guises. If humanism was partly about an emergent 'human' based upon notions of individual autonomy and the rational mastery of nature, broadly construed to include one's embodiment, we can in one sense partly see transhumanism as a form of hyper-humanism (see Thacker, 2003: 75). In fact within limits I argue that transhumanism has links to both humanism *and* critical posthumanisms. Bostrum, one of the leading proponents of transhumanism, defines it as

> . . . a loosely defined movement that has developed gradually over the past two decades, and can be viewed as an outgrowth of secular humanism and the Enlightenment. It holds that current human nature is improvable through the use of applied science and other rational methods, which may make it possible to increase human health-span, extend our intellectual and physical capacities, and give us increased control over our own mental states and moods (2005: 202/3).

Transhumanists take things literally. Their supersession of humanism is material in a specific way. When they talk of posthumans they are imagining a human materially modified, a body 'enhanced'.[17] This is hyper-humanism in the sense of bodily and emotional control; and in the stress upon individual autonomous choice over current and forecasted reproductive technologies (see Bostrum, 2003; Savulescu, 2001). The emphasis on the individual here is counter to anti-humanist critiques of liberal humanist thought and the extension of the value of control is counter to critical posthumanism. Although it remains faithful to a human both individualised and rational, it extends these both, for example, through a discourse of individual freedom to 'biologically enhance' oneself or one's offspring (Bostrum, 2004), and the rationalization of the body through genetic manipulation and 'optimization'. David Pearce, founder of the World Transhumanist Association with Bostrum, expresses a techno-utilitarianism in his wish to use biotechnology to 'abolish suffering in all sentient life' (2007: 1). Unlike Bostrum, he extends his thesis to include the nonhuman describing his 'abolitionist'[18] view as 'Bentham plus biotechnology'. Like critical posthumanism he disavows anthropocentric bias, but he advocates technological solutions to the industrialized commodification of animals such as the development of in-vitro meat and the possible technological redesign of the global ecosystem[19] or the 're-programming' of vertebrate genomes to alleviate suffering. A further recent inclusion of nonhuman animals within transhumanism can be found in the techno-salvational notion of 'uplift' (Dvorsky, 2008). This argues that there is an ethical imperative to use technology to enhance the intelligence not only of ourselves but that of nonhuman animals. The stark paternalistic anthropocentrism of this idea can be gleaned from the very term 'uplift'. These are idiosyncratic and rare mentions of animals and 'nature' in transhumanist discourse. In their ambivalence to anthropocentrism they are further evidence of the complexity of the broader posthumanist terrain.

Whilst such views may irk or confuse many environmentalists and animal advocates,[20] arguably the initial offence caused by transhumanism, at least to those of a humanist persuasion, is found within its radical critique of ideas of human essence. This kind of objection is illustrated by Francis Fukuyama (2004) who chose transhumanism in a special series in the journal *Foreign Policy* on the World's Most Dangerous Ideas. Specifically he frames transhumanism as a movement that seeks to modify 'human essence' and is therefore something which, for him, potentially threatens liberal democratic citizenship,

> Underlying this idea of the equality of rights is the belief that we all possess a human essence that dwarfs manifest differences in skin colour, beauty, and even intelligence. This essence, and the view that individuals therefore have inherent value, is at the heart of political liberalism (2004: 42).

Such an objection stems from a view of the 'human', not as emergent, but as both stable and knowable. Unsurprisingly Bostrum (2004) takes issue with Fukuyama's notion of human essence; although interestingly by recourse to the science of evolutionary biology which he argues shows that our gene pool is in 'constant flux'. It is more or less predictable that Bostrum does not try to undermine Fukuyama's notion of human essence with the point of 'rationality' and 'autonomy' being historically contingent markers of the human as this would be to undermine his own transhumanism which valorizes these same markers. It is at this point, however, that we can say that both transhumanism and critical posthumanisms are similar, at least in the sense of being critical of notions of human essence.[21]

c) *Critical posthumanisms as a critique of transhumanism*

The critical posthumanist response to transhumanism is probably best summed up in terms of exasperation. If the reason for this feeling can be pinpointed, it is to the failure of transhumanism to engage critically with the constellation of dualistic ontology which has fuelled humanism's capacity for exclusion and othering. Consequently it can at times appear both misanthropic and somatophobic wherein the active scientific *mind* is continually re/cast to work to improve the always inferior passive *body*. Transhumanism can consequently come over as intolerant to various forms of embodied difference, notably aged and disabled bodies. Although not simplistically a reproduction of eugenics, it attempts to convert that movement into an acceptable version through discourses of individual consumer free choice and responsibility. The transhumanist unease with death, where the ageing process is redefined as pathology, is characteristic of a movement unreconciled with the ecological situatedness of the human, and determined to broaden further the modernist notion of human freedom as an escape from 'nature'. Transhumanism also mimics popular science through an over-confident attachment to genetic determinism, professing already the wide range of behaviours and physicalities that will be genetically controllable; doubly problematic, since many of these, such as 'intelligence' and 'beauty', are anyhow conceptually nebulous. Thus Thacker's critique of transhumanism as a movement that has a tendency toward scientism, 'to apply life science concepts toward social and political problematics' (2003: 75). As a critical posthumanist Thacker is also questioning of the transhumanist conceptualization of information and its ontology of technology. He targets the transhumanist idea of the material world simply as information, uploadable and editable as if computer code which is a partial account of (for example) the body, effacing many other senses in which the body may be conceived. This echoes Hayles' (who he cites) work on cybernetics who argues when '..the posthuman constructs embodiment as an instantiation of thought/information, it continues the liberal tradition rather than disrupts it' (1999: 5). Thacker's point on ontology is also an important one, especially for thinking through the differences between transhumanism and other posthumanisms. He makes the point that transhumanism relies upon a construction of technology as a tool arguing that 'the blind spot of this thread of posthumanism is that the ways in which technologies are themselves actively involved in shaping the world are not considered' (2003: 76). This is contra a central tenet of science studies and its commitment to an antihumanist ontology of nonhuman agency, a counter to human over-confidence in the control of technology. Thacker makes the point that it is precisely the granting of agency to the human and passivity to technology – a dualistic ontology of human and machine – that is crucial to transhumanist fantasies (Thacker, 2003: 77).

Thacker's critique of transhumanist takes on embodiment highlights an important schism between transhumanism and critical posthumanisms. Such approaches attempt to render materiality economically intelligible helping to interweave science and capital, promoting new accumulation strategies. Just as the self-possessive individualism of the liberal humanist subject is partly produced by market relations and does not predate them (Hayles, 1999: 3), transhumanism emerges at a later stage of capitalism when the focus of the eye of accumulation has turned molecular. Critical posthumanisms tend on the other hand to critique capitalist development and its intersection with a particularly hierarchical discourse of species. Thus in the broader field of animal studies the history of capitalism may be told in terms of particular human/animal relations. Or in terms of human 'enhancement', critical work may centre on the emergence of parenthood and bio-power (Graham, 2002: 111) or what new markets for genetic testing or pre-implantation genetic diagnosis (PGD) may be saying about the 'human'. This is not to say that there is necessarily a coherent critique of capitalism yet emerging from critical posthumanisms but that its critical moves in this respect mark it out from transhumanist naturalizations of market norms.

Critical posthumanisms – a difference of emphasis

As we saw above, there are certainly close affinities between antihumanism and critical posthumanisms; and even some aspects in common between transhumanism and critical posthumanisms. There are also, to recall, stated commonalities, especially in the work of Hayles (1999, 2003) and Badmington (2003) between humanism and critical posthumanism. Badmington is not concerned with an absolute break with humanism, rather he writes, 'If the version of posthumanism that I am trying to develop here repeats humanism, it does so in a certain way and with a view to the deconstruction of anthropocentric thought' (p. 15). Badmington is correct to argue against the posthuman as necessarily a temporal epoch, and rather, as I suggested above, some of the tenets of critical posthumanism could in fact be seen as co-existent and interior to humanism. In spite of these connections within the terrain there are articulated differences. However, we should resist a simple dualism between transhumanism and critical posthumanism given the range of positions which fall under these terms. As illustrated, within transhumanism there are different degrees of position.

But now I explore again, in more detail, a difference of emphasis within the terrain of critical posthumanisms. There is a disconnect between some critical posthumanist writers, largely to be found within science studies, and the substantial body of work of those writing in animal studies and environmental ethics who are doing significant critical posthumanism.[22] The gap is not total but revolves around different uses of the word anthropocentrism which may be used either in a theoretical ontological sense and/or a more political sense related to critiques of human supremacism. I want to argue that critical posthumanism is at its most coherent and incisive when it builds upon *both* these uses, seen notably in the work of Wolfe (2003), Haraway (eg 2008), Rossini (2006) and Hayles (1999). For example Wolfe and Haraway have longstanding theoretical and political interests in human/animal relations, whereas the main focus of Hayles has been on cybernetics and embodiment, including contributions to counter dualistic ontology. As she writes:

> . . . my dream is a version of the posthuman that embraces the possibilities of information technologies without being seduced by fantasies of unlimited power and disembodied immortality, that recognizes and celebrates finitude as a condition of human being, and that understands human life is embedded in a material world of great complexity, one on which we depend for our continued survival (1999: 5).

Again this is said in contrast to transhumanist fantasies but here also with an explicit ecological frame. Arguably it's this, the questioning of human hubris, which gives critical posthumanism a purpose and resonance with contemporary concerns over society-nature conceptions and inter-relations. Otherwise it risks appearing indistinct from the considerable body of critical scholarly work on dualistic concepts (see Castree and Nash, 2004: 1342). Ontological reconstruction work conjoined to an ecological frame remains suspicious of the 'human' yet also wishes to construct a human with moral import and relationality. 'Nature' here is not only agential and interwoven with 'culture', it is also ethically significant. Such a critical posthumanism is critical of science yet not anti-science. It is in the ongoing formulation of this problematic – one which draws lines of connectivity between and within the 'human' and 'posthuman' – that the most promising critical posthumanism is to be found. However neither the moral decentring of the human nor the politicization of the nonhuman here is in anyway simple. For example, many, though by no means all, accounts of animal ethics reproduce humanist notions of essence in the space of the nonhuman by using anthropocentric criteria of value (Wolfe, 2003: 10). Similarly ecological arguments may project anthropocentric and humanist desires for 'equality, respect and freedom' onto 'nature'[23] or indeed the notion of an 'environmental' ethics may be indulging in the mirror opposite ontological purity game that has excluded the nonhuman from the social (see Castree, 2003b).

This disconnect between these two strands of critical posthumanism has arisen partly because the main focus of many science studies scholars has been upon disciplinary critique and ontological creativity to challenge core concepts in the social sciences, whereas for many writing in environmental and animal studies there has been a more explicit normative emphasis to their posthumanism. The former sense is captured well in a recent edited collection coming out of actor-network theory (ANT) entitled *Material Agency – Towards a Non-Anthropocentric Approach* (Knappett and Malafouris, 2008). Here 'anthropocentric' is primarily an ontological rather than an ethical signifier, and so used differently to its use in, for example, animal studies, environmental ethics, rural sociology and human geography.

However this is another fuzzy distinction which may be probed a little further by reference to a paper by Law and Mol (2008) in the aforementioned collection. Writing on the UK Foot and Mouth disease outbreak in 2001 they consider the different kinds of sheep enacted by ovine enmeshments in various scientific and economic knowledge claims as well as the sorts of ways in which sheep made a difference to human conceptualizations of them and the trajectory of the disease. In common with much ANT the paper seeks to queer notions of control, agency and the actor. It takes sheep seriously in a sociological sense, resisting their casting as passive recipients of knowledge and instrumentalization. In such a framing it may be said that the ontological and the ethical are not wholly distinct. ANT does make a serious attempt to conceive of animals as parts of agential networks which challenges traditional social science thinking on agency. Clearly this is not the same as normative work on human/animal relations; but there is an overlap here between the ontological and the normative, since accounts of human/animal and human/nature relationalities that intend to decentre the human in more explicitly political ways also require some way of challenging traditional objectifications of the nonhuman.[24] In other words the ontological re-framing makes the nonhuman matter. Even if in the perhaps limited sense of being granted admission to the social it can still be cast as a move that is of importance to normative reconceptualizations as it begins to try to unravel the various interlinked hierarchical dualisms that have solidified notions of essence and power. This, then, is an area of overlap and tension within the space of critical posthumanism, one which speaks to broader debates on the relationship between ontology and ethics, debates around the so-called 'naturalistic fallacy' and between conflicting conceptions of, and admissions to, the political.

Posthumanisms and/in genomics

This discussion brings home the importance of ontological re-framing in the varying strands of posthumanist thought, moves which emanate from the social sciences, the humanities *and* the natural sciences. As we saw earlier, recent evolutionary theory has witnessed the ability of genomics to play a role in such ontological re-framing. In the final section of this chapter I consider the sort of ontological moves in process within genomics and related bio-sciences. To what extent do they resonate with strands of posthumanist discourse? What are the intimations of 'nature' in genomics, and what ethics are at play?

Initially, it is important to underline that there is a lot more going on in contemporary (biological) science than genomics. For example one can make the general point that animal welfare and behavioural sciences are engaged in bringing forth richer accounts of animal subjectivities which disrupt the restriction of 'mind' to the human. Moreover whilst I have already shown how genomics may be acting to redefine theories of human evolution, evolutionary theories themselves are in flux. Notably the work of biologist Lynn Margulis (eg 1981) has been taken up by social scientists to re-orient assumed anthropocentrism; for example, by decentring *sexual* reproduction (Hird, 2006) and by underlining the importance of boundary subverting microbial symbiotic relations to (human) evolution (see Parisi, 2007: 31/2). The philosophical work of Dupré and O'Malley on metagenomics[25] also thinks through the ontological ramifications of taking microbes seriously (2007). Such work, which also weakens *disciplinary* boundaries, nourishes a posthuman ontology that attends to interconnection, interdependency, and relationality (eg Castree, 2003a); hybridity (Whatmore, 2002) and process (Braidotti, 2006; Twine, 2001; Abram, 1996; Young, 1990). However genomics and related bio-sciences can be seen to take quite heterogeneous paths in relation to such ontological re-framings and posthumanist discourse.

Indeed, initial received wisdom held that genomics reinscribed a notion of essence around an over-valorized 'gene' which would lend itself to narratives of isolation, manipulation and control. It would be fair to say that this frame persists but that it has been muddied by a shift to complexity in the context of post-genomic research. The aforementioned narratives may be challenged by the appearance of new puzzling actors in the shape of complex gene or genome interactions, or epigenetics. Such indications of complexity can be seen as correctives to scientific boundary making and complementary to the sort of anti-dualistic ontologies of critical posthumanism. The extent to which we may characterize an alignment between genomics and various strands of posthumanism is linked strongly to our liberal working definition of 'genomics'. It is important then to be clear that the conservative definition of genomics, used for example when talking about techniques of plant and animal breeding,[26] is being bracketed here in order to give a fuller sense of the connections between contemporary genetics and other areas of science such as stem cell research and reproductive science generally. Taking this broader view arguably highlights the heterogeneity of genomic natures.

Genomics is apparently couched in humanism. It promises self-knowledge and to cure human dis-ease. Comparative genomics uses other species to find out more about 'ourselves'. Experimental transgenics expands the number of nonhuman animals used in medical experimentation. It 'humanizes' them to make them better models. As mentioned earlier it hopes to unravel the 'something special' of the 'human'. It becomes *trans*humanist when it is re-constructive; as in the creation of patentable transgenic animals or in the burgeoning field of synthetic biology. Genomics could then be considered transhumanist both in the sense of extended human mastery of the nonhuman and in the promise of a future altered human.[27] Several examples of work on nonhuman animals could be seen in this light. We have already heard mediated stories of 'smart mice', 'Schwarzenegger mice' and 'fearless mice' (see Twine, 2007: 514); enhanced animals for an unclear, hypothetical enhanced human future. But these re-constructions also call into question a simplistic attachment of genomics to humanism. The question of what sort of post/humanist discourse genomics might be said to embody echoes the complexity and relationality of the posthuman terrain itself. In many of the biotechnological re-constructions that we now see it is relevant that the human and nonhuman are either juxtaposed, or brought into close proximity, or indeed co-mingled. It is perhaps this point that begins to facilitate the notion of a humanist or transhumanist genomics as contradictory. I focus on this by drawing upon two examples of human/non-human re-construction: the use of interspecies databases in farm animal comparative genomics, but first the question of hybrid embryos in stem cell research.

The use of nonhuman animal eggs to make human/nonhuman hybrid embryos can only occur in the first instance due to the construction of animal material as resource, an unapologetically anthropocentric frame. The industrialized breeding of various species provides the standing reserve, if required, of nonhuman body parts for human medical research. Bovine egg sources[28] were turned to after stem cell scientists realized that human female sources were in short supply and involved surgical and drug intervention. From the outset, then, hybrid embryos are built upon a traditional human/nonhuman ethical divide. So far so humanist. Moreover, the UK's Human Fertilisation and Embryology Authority deemed that hybrid embryos should be regarded as human in spite of the presence of nonhuman mitochondrial DNA. The material structure and constitution of the hybrid embryo could also be seen to map onto traditional dualistic associations of materiality and human/animal difference. Historically the animal has been divided off from the human, a distinction which has mapped onto that between body and mind. Mindlike qualities have been seen as owned by the human, whereas the animal has been represented as physicality and instinct. The notion of the human/animal hybrid embryo could then be said to embody this association where the animal is literally the shell, the body, the outside, the background, whereas the human is the nucleus, the centre, the mind, and the locus of 'real' agency. The marginalization of the animal here is reflected in both the very materiality of the hybrid embryo as well as in the arbitrariness of its construction as effectively 'human'.

Such definitions arguably attempt to disguise the proximal interspecies relations at play here. Proximity enables hybrid embryos to act as potential or perceived identity threats, as suggested by public perception research (Parry and Faulkner, 2007). Why might hybrid embryos be construed as offensive to the human, to a notion of 'human dignity'? Boundary crossings may be seen as a breach of essentialized categories. More importantly we live in a cultural context which still speaks with varied discourses of animality, whereby 'animal' is often a pollutive and degrading signifier, and is able to be so because of its designated inferior position vis-à-vis the human. One kind of yuck factor response may speak to a fear that the material mixings of human and animal are accompanied by the symbolics of animality breaching and 'degrading' the human. This would seem to be the case for some of the respondents in Parry and Faulkner's research, expressing fears that hybrid embryos could 'undermine the specialness of humans'. Hybrids produce reflexivity on human/animal relations, differences and similarities. Some interviewees spoke of 'consciousness' and 'sentience' as signifiers of the human, but as many animal scientists now argue (and EU law in the case of 'sentience'), these properties are not now seen as unique to the human.

Hybrid embryos are also subject to the 14-day rule familiar in human embryonic stem cell research. Obviously nonhuman animal embryos would not normally be subject to this. In other words it is a human criterion in spite of the observation that the development of the so-called 'primitive streak' is something that we share with other mammals, birds and reptiles. This appears at different times according to a given species' gestation period however it is not in fact a particularly human developmental characteristic that continues to be used as a moral boundary. In some instances hybrid embryos are being offered more regulatory protection than human embryos. As Brown has remarked, a ban on creating hybrid embryos for research but not human embryos 'places humans in the morally awkward position of being experimental subjects in a way that transpecies (less than human) embryos are not' (2009: 154). One could ponder whether hybrid embryos destabilize human/nonhuman hierarchies, or whether they reinforce them, or whether in fact they could be seen to do both these things in various contexts. It should be noted that eventually the UK government, after pressure from stem cell scientists, reversed their earlier ban giving the green light to hybrid embryo research. Although one must concur with Brown that such chimeric actors hardly signal a collapse of humanist hierarchies (2009: 162), they minimally have the capacity to bring into relief some of the arbitrary conventions related to the anthropocentrism of hyper-separating 'human' and 'animal'.[29] In this vein they may be said to point toward a critical posthumanism, even as they concurrently remain within a humanist medical trajectory.

The use of interspecies databases in farm animal comparative genomics provides another example of the 'human' and animal' brought into proximity, albeit in a quite different sense. Here proximity is virtual but no less material. The conversion of animal bodies into mediated virtual representations of genomic data is now the norm for scientists working in animal production. The genomic sequencing of animals used in agricultural production has produced a wealth of databases where chromosomes can be searched for 'economically useful' coding genes related to a whole range of characteristics such as muscling, disease resistance and meat quality. Initially the *distancing* and fragmenting quality of this can be highlighted. Scientific practice, removed from the sensual presence of animal bodies, echoes the distance achieved for and by the consumer in the conversion of animals into the category of meat. Here the mediation is primarily productive, fitting unproblematically into a humanist narrative of controlling, in ever nuanced ways, the nonhuman. However *proximity* is also productive in the scientific practice of comparative genomics. Degrees of shared ancestry between species give rise to homologous genes forming 'multi-gene families' and shed new light on species relationality and evolutionary change. But the primary focus in agricultural science is on comparative mapping between different species in order to secure novel genetic intervention. *This means bringing the sequenced human genome into the realm of meat production.*

Using human genome sequenced data comparatively, as a railroad into the new frontier of farm animal genomics, takes place in a variety of ways. In a clever pun[30] Meyers *et al.* (2005) in their paper '*Piggy-BACing the Human Genome II. A High Resolution, Physically Anchored, Comparative Map of the Porcine Autosomes*' comparatively map this new territory in a more targeted way than had previously been possible in order to glean more about economically relevant areas of the porcine genome. In the search for new forms of capitalization, mammalian relationality forms a basis to the relevance of human genomic data.[31] Grapes *et al.* (2005) have also commented upon the usefulness of the human genome to porcine genomics, specifically in relation to a high human-pig correlation in comparing mutation number in genes. Research into other major agricultural species such as bovine (eg Everts-van der Wind *et al.*, 2004) and ovine (eg Wu *et al.*, 2008) genomes also draws upon human genomic data. Geneticist James E. Womack has described the Human Genome Project as opening the door to farm animal genomics but said that in turn this work informs the human genome (2005: 1699). This latter aspect is expressed in stronger terms by de Koning *et al.*, asserting that farmed species can bridge the research model gap between 'mice' and 'men' because 'their biology is often much closer to that of humans' (2008: 483). This would suggest the capacity of genomics to alter the meanings of different animals, in this case an instrumental 'doubling up' of farm animals as both food *and* medical research model.

Again in certain ways this is classically anthropocentric; a clustering of novel use-values for the nonhuman but in this case arguably also a new legitimation frame for animal science.[32] The co-mingled re-constructive possibilities for human and animal also place such comparative genomics work in the terrain of transhumanism. In the language of farm animal genomics, these possibilities are consistently coined in terms of the search for 'economically interesting' traits, illustrating not merely the alliance of this type of animal science to commerce, but more fundamentally the location of animal science *within* the economic and moral spheres of constructing human/animal futures. In human/ist terms possibilities are typically stressed in terms of health and the alleviation of suffering but it is worth noting that animal genomic interventions are also framed in terms of a contribution to animal health, and justifications for funding human genomics are also framed in terms of economic gain. Not only does comparative genomics work bring 'human' and 'animal' into close *proximity* and confuses notions of what may be deemed 'human DNA' or 'animal DNA', it also contributes to a discourse of human/animal *similarity*. Thus we may note one sense in which such animal scientists are distinctly *non*-anthropocentric. Both humans and animals are subject to what Thacker has termed 'informatic essentialism' which proposes that

> . . . the relationships between the biological body and information technology is such that the body may be approached through the lens of information . . . , is therefore subject to the same set of technical actions and regulations as is all information. In short, when the body is considered as essentially information, this opens onto the possibility that the body may also be programmed and reprogrammed (2003: 86).

Although Thacker here is describing a logic of transhumanist 'enhancement' we can also tease out the destabilizing element from comparative genomics. Introducing the 'human' into the realm of farm animal genomics and vice versa erodes human/animal dualism and implicitly questions the subtext of anthropocentrism. By introducing comparative interspecies experimental modelling into farm animal science, meat production (in a material-semiotic sense) potentially destabilizes itself, in that it challenges the claim of difference so germane to the legitimation of the human consumption of other animals. In common with hybrid embryos the use of interspecies databases in farm animal comparative genomics constitutes a further way in which genomics embodies humanist values. It is also contradictory, however, and potentially points to more critically posthumanist directions of thought, strands that, in this case, specifically question the naturalization and depoliticization of human/animal relations.

Conclusions

This chapter has outlined a narrative of the posthuman terrain, highlighting various interconnections and specificities within. For example, both transhumanist and critical posthumanist strands of thought point to nature/culture hybridity, yet with the former conceiving of this as a relatively recent technologically mediated development, and the latter wanting to engage in an ontological politics which says 'we have always been hybrid'. Indeed critical posthumanism might be summed up in its approach to hybridity: as a refutation of the charge of 'unnaturalness' *but* a cautious sensitivity toward moves of naturalization. Varied deployments of genomics, like wider biological science, have a complex relation to 'nature', and so also the 'human'. Work on human evolution, microbial genomes and genomic interaction intersects with accounts of human/nature inter-relationship, questioning the purity of the 'human' and the separability of 'nature'. Moreover the human genome project itself delivered an initial jolt to hubris in its revelation of a much lower than expected number of human genes vis-à-vis other species. Other areas as the examples above suggest seem to indicate a continuation or even extension of humanist values even as they risk instability through interspecies proximity. If such 'perfecting' moves of transhumanism can be said to bring into relief the religious residues and market efficiency drives of the humanist project, then it may be that a critical posthumanism, which labours to dispute both ontological purity and anthropocentrism, signposts the vital conceptual shifts for rethinking the social and political complexities of interspecies relationality.

Acknowledgements

I thank John Dupré, Helena Pedersen, Tim Newton and Gail Davies for very useful comments. The support of the Economic and Social Research Council (ESRC) is gratefully acknowledged. The work was part of the programme of the ESRC Centre for Economic and Social Aspects of Genomics (Cesagen).

Notes

1 It is important to appreciate that if the Human Genome Project can confidently distil the human, it is also involved in a project to say what the nonhuman is not.
2 I should disclaim some historical generality here. I accept the argument that it is misleading to overstate that a particular epoch is dominated by any one model of nature, and that important anti-dualistic forms of thought were similarly important during the Enlightenment period (Newton, 2007: 37; see also Meyer, 1999).
3 I will consider the concept in much greater detail later on.
4 I take the point from Science and Technology Studies scholars (eg Latour, 2000: 114) that words such as 'mastery' are problematic in that they appear to disallow for both recalcitrance and agency from nonhuman nature. Moreover it could be said to portray 'nature' as something always *acted upon* by the 'human'. Mindful of these points, but also in light of ongoing ecological crises, the animal-industrial complex and the deleterious effect upon human health of both, I remain strategically content with 'mastery'.
5 As will become clear, I mean this in the sense used by both scholars writing in environmental and animal ethics, as a conception that is critical of the human mastery of an externalized nature.
6 See [http://news.bbc.co.uk/1/hi/sci/tech/7570291.stm]. It's not clear whether such tests are supposed to imply a degree of species superiority for those skilled enough to pass, or whether the tests are not themselves anthropocentric in their conception of 'skill' accessed 20 August 2009.
7 There is still some debate over whether the recently discovered *Homo floresiensis* actually constitutes a distinct species.
8 See http://en.wikipedia.org/wiki/Neanderthal_Genome_Project as well as, more generally, the Genographic Project https://www3.nationalgeographic.com/genographic accessed 20 August 2009.
9 Early reference to 'antihumanism' is found in Althusser's interpretation of a shift to theoretical antihumanism in Marxist theory (see Althusser, 1966: 219–47; and Davies, 1997: 57–60).
10 Although I am arguing here, in similarity with Braidotti (see quote p. 179), for a difference within the 'posthuman' between antihumanism and critical posthumanism, they are obviously overlapping. Much theory calling itself posthumanist also carries with it the arguments of antihumanism and some theorists, for example Heidegger, can be seen as important bridges between antihumanism and early intimations of an ecologically inflected critical posthumanism.
11 I define feminism broadly here to include intersectionality with class, race, age and disability which all surely inform a hegemonic 'human'.
12 Certainly many of the ontological points of contemporary posthumanism owe an ancestral debt to their coverage in earlier debates within feminist theory, see, for example, Stanley and Wise (1983).
13 Plumwood actually refers to anthropocentrism as 'anthrocentrism' as a representational means in her analysis to signal its affinity with *andro*centrism.
14 By using the term ecological throughout I should not be simply read as prioritizing 'environmental' ethics over ethical relations with other animal species.
15 Sloterdijk is an interesting figure in the sense that he unusually combines the sort of ontological work associated with critical posthumanisms and antihumanism, together with more transhumanist values (see Sloterdijk, 2009).
16 To focus instead upon 'a discourse of species' is to think through the histories and politics of species relationality, to reflect critically, for example, upon their humanist inflection (see Wolfe, 2003: 2).
17 This is also seen on the main transhumanist web-site where it is confidently stated that 'At present, there is no manner by which any human can become a posthuman' (see http://www.transhumanism.org/index.php/WTA/faq21/90/) accessed 20 August 2009. Such a view is dependent upon a literal interpretation of 'being posthuman'. Yet if we define 'being posthuman' in the sense which some critical posthumanists mean, then it is surely possibly to be posthuman;

with vegan feminists being a good example. Veganism radically undermines anthropocentrism *and* materialises a different human (body).

18 This refers to the abolition of suffering, and should not be confused with the application of the term to animal rights advocates who mean an abolition of the view of animals as property (see Francione, 1996).

19 Presumably this is meant in contrast in some way to the re-design already well underway. Yet this does resonate with contemporary ideas on geo-engineering solutions to climate change.

20 I am making the assumption that most environmentalists would not favour such a techno-managerialist response, and that most animal advocates would not take such an extreme view on suffering or predation.

21 Moreover both transhumanism and critical posthumanisms, despite quite profound differences, share an interest in the hybridization of the human and nonhuman.

22 The work of Lynda Birke, Donna Haraway and Katherine Hayles are important bridging points here.

23 I thank Tim Newton for this point.

24 One might counter against this view of a potential affinity between ontological antihumanists such as ANT scholars and ethical posthumanists by arguing that theories such as ANT may not satisfy the focus upon *individual* animals that many animal ethics demand.

25 Dupré and O'Malley define metagenomics as '*the genome based analysis of entire communities of complexly interacting organisms in diverse ecological contexts*' (2007: 835).

26 Part of the reason for insisting on this distinction has emanated from the politics of genetic modification (GM) and the desire to not see genomics (in this narrow sense) tainted by GM. Although the practice of GM is more manipulative vis-à-vis genomics, the development of GM cannot be said to be separate from the development of genomic sequencing.

27 It should be added that a wholesale denigration of the transhuman is problematic. It ignores how new technologies can shift the terms of choice and ethics. Thus pre-implantation genetic diagnosis (PGD) enables 'suspect embryos' to be screened for serious genetic conditions. The politics over designer babies takes place over contestations of the word 'serious'.

28 Rabbits and pigs are also used.

29 For a discussion on dualism and the concept of hyper-separation, see Plumwood (1993).

30 BAC stands for bacterial artificial chromosome, an artificially constructed segment of nucleic acid often used in genome sequencing projects.

31 Mice, rapidly sequenced owing to their use as a model in medical research, represent a further 'non-agricultural animal' used as a model genome in comparative agricultural research.

32 Productivist animal science is unsurprisingly considerably less funded than science related to human health. Therefore it is clearly to its benefit to stress its relevance to human health.

References

Abram, D., (1996), *The Spell of the Sensuous: Perception and Language in a More-Than-Human World*, London: Vintage.

Althusser, L., (1966), *For Marx*, Harmondsworth: Penguin.

Badmington, N., (ed.), (2000), *Posthumanism*, NYC: Palgrave.

Badmington, N., (2003), 'Theorizing Posthumanism', *Cultural Critique*, 53: 10–27.

Bostrum, N., (2003), 'Human Genetic Enhancements: A transhumanist perspective', *Journal of Value Inquiry*, Vol. 37(4): 493–506.

Bostrum, N., (2004), 'Transhumanism: The world's most dangerous idea?' available online at http://www.nickbostrum.com/papers/dangerous.html (last accessed October 2008).

Bostrum, N., (2005), 'In Defense of Posthuman Dignity', Bioethics Vol. 19(3): 202–214.

Braidotti, R., (2006), 'Posthuman, all too human – Towards a Process Ontology', *Theory, Culture and Society* ,Vol. 23(7–8): 197–208.

Brown, N., (2009), 'Beasting the Embryo: the metrics of humanness in the transpecies embryo debate', *Biosocieties*, Vol. 4 (2–3): 147–163.

Butler, J., (2004), *Undoing Gender*, London: Routledge.

Castree, N., (2003a), 'Environmental issues: relational ontologies and hybrid politics', *Progress in Human Geography*, Vol. 27(2): 203–211.

Castree, N., (2003b), 'A Post-environmental Ethics?' *Ethics, Place and Environment*, Vol. 6(1): 3–12.

Castree, N. and C. Nash, (2004), 'Introduction: posthumanism in question', *Environment and Planning A*, Vol. 36: 1341–1343.

Castree, N. and C. Nash, (2006), 'Editorial – Posthuman Geographies', *Social and Cultural Geography*, Vol. 7(4): 501–504.

Davies, T., (1997), *Humanism*, London: Routledge.

de Koning, A. Archibald and C. Haley, (2008), 'Livestock Genomics: bridging the gap between mice and men', *Trends in Biotechnology*, Vol. 25 (11): 483–489.

Derrida, J., (2003), 'And say the animal responded?' in Wolfe, C. (ed.), *Zoontologies: The Question of the Animal*, Minneapolis and London: University of Minnesota Press: 121–46.

Didur, J., (2003), 'Re-embodying Technoscientific Fantasies – Posthumanism, Genetically Modified Foods, and the Colonization of Life', *Cultural Critique*, Vol. 53: 98–115.

Dupré, J. and M. O'Malley, (2007), 'Metagenomics and biological ontology', *Studies in History and Philosophy of Biological and Biomedical Sciences*, Vol. 38: 834–846.

Dvorsky, G., (2008), 'All Together Now: Developmental and ethical considerations for biologically uplifting nonhuman animals', *Journal of Evolution and Technology*, Vol. 18(1): 129–142.

Ehrenfeld , D., (1981), *The Arrogance of Humanism*, Oxford: Oxford University Press.

Eswaran, V., H. Harpending and A.R. Rogers, (2005), 'Genomics refutes an exclusively African origin of humans', *Journal of Human Evolution*, Vol. 49: 1–18.

Everts-van der Wind, A., S.R. Kata, M.R. Band, M. Rebeiz, D.M. Larkin, R.E. Everts, C. Green, L. Liu, S. Natarajan, T. Goldammer, J.H. Lee, S McKay, J.E. Womack and H.A. Lewin, (2004), 'A 1463 Gene Cattle-Human Comparative map with anchor points defined by human genome sequence coordinates', *Genome Research*, Vol. 14: 1424–1437.

Francione, G., (1996), *Rain Without Thunder: the ideology of the animal rights movement*, Philadelphia: Temple University Press.

Fukuyama, F., (2004), 'The World's Most Dangerous Ideas – Transhumanism', *Foreign Policy*, September/October: 42–43.

Gane, N., (2006), 'When we have never been Human, what is to be done? Interview with Donna Haraway', *Theory, Culture and Society*, Vol. 23(7–8): 135–158.

Garrigan, D. and M.F. Hammer, (2006), 'Reconstructing human origins in the genomic era', *Nature Reviews Genetics*, Vol. 7: 669–680.

Graham, E.L., (2002), *Representations of the Post/Human – Monsters, Aliens and Others in Popular Culture*, Manchester: Manchester University Press.

Grapes, L., S. Rudd, R.L. Fernando, K. Megy, D. Rocha and M. Rothschild, (2005), 'Searching for Mutations in Pigs using the Human Genome', *Iowa State University Animal Industry Report* Available online at http://www.ans.iastate.edu/report/air/2005pdf/2024.pdf (accessed October 2008).

Haraway, D., (1997), *Modest Witness@Second Millenium. FemaleMan Meets OncoMouse™: Feminism and Technoscience*, London: Routledge.

Haraway, D., (2008), *When Species Meet*, London: University of Minnesota Press.

Hayles, N.K., (1999), *How we became Posthuman – Virtual bodies in Cybernetics, Literature, and Informatics*, Chicago: University of Chicago Press.

Hayles, N.K., (2003), 'Afterword: The Human in the Posthuman', *Cultural Critique*, 53: 134–137.

Hird, M., (2006), 'Animal Transex', *Australian Feminist Studies*, Vol. 21 (49): 35–50.

Knappett, C. and L. Malafouris, (eds.), (2008), *Material Agency: Towards a Non-Anthropocentric Approach*, New York City: Springer.

Latour, B., (2000), 'When things strike back: a possible contribution of 'science studies' to the social sciences', *British Journal of Sociology*, Vol. 51(1): 107–123.
Law, J. and A. Mol, (2008), 'The Actor-Enacted: Cumbrian Sheep in 2001', in Knappett, C. and L. Malafouris (eds), *Material Agency: Towards a Non-Anthropocentric Approach*, New York City: Springer.
Margulis, L., (1981), *Symbiosis in Cell Evolution – Microbial Communities in the Archean and Proterozoic Eons*, (2nd edn.), New York: W.H. Freeman and Company.
Meyer, J.M., (1999), 'Interpreting nature and politics in the history of Western thought: The environmentalist challenge', *Environmental Politics*, Vol. 8(2): 1–23.
Meyers, S.N., M.B. Rogatcheva, D.M. Larkin, M. Yerle, D. Milan, R. Hawken, L.B. Schook and J.E. Beever, (2005), 'Piggy-BACing the human genome II. A high resolution, physically anchored, comparative map of the porcine autosomes', *Genomics*, Vol. 86: 739–752.
Newton, T., (2007), *Nature and Sociology*, London: Routledge.
Parisi, L., (2007), 'Biotech: Life by Contagion', *Theory, Culture and Society*, Vol. 24(6): 29–52.
Parry, S. and W. Faulkner, (2007), 'Talking about hybrid embryos for stem cell research', paper presented at a *Symposium on the The UK Transpecies Embryo Debate*, ESRC Genomics Forum, University of Edinburgh, 27th November 2007.
Pearce, D., (2007), 'The Abolitionist Project' available online at http://www.hedweb.com/abolitionist-project/index.html (last accessed October 2008)
Plumwood, V., (1993), *Feminism and the Mastery of Nature*, London: Routledge.
Plumwood, V., (1996), 'Androcentrism and Anthrocentrism: Parallels and Politics', *Ethics and the Environment*, Vol. 1(2): 119–152.
Rawles, K., (2007), 'Animals versus the Environment – Is animal welfare a luxury in the fight against climate change?' *Food Ethics*, Vol. 2(4): 13–15.
Rossini, M., (2006), 'To the Dogs: Companion speciesism and the new feminist materialism', *Kritikos*, Vol. 3 http://intertheory.org/rossini
Savulescu, J., (2001), 'Procreative Beneficence: Why We Should Select the Best Children', *Bioethics*, Vol. 15(5–6): 413–426.
Sloterdijk, P., (2009), 'Rules for the Human Zoo: a response to the Letter on Humanism', *Environment and Planning D: Society and Space*, Vol. 27(1): 12–28.
Stanley, L. and S. Wise, (1983), *Breaking Out Again – Feminist Ontology and Epistemology*, London: Routledge.
Thacker, E., (2003), 'Data made flesh: Biotechnology and the discourse of the posthuman', *Cultural Critique*, 53: 72–97.
Twine, R., (2001), 'Ma(r)king Essence-Ecofeminism and Embodiment' *Ethics and the Environment*, Vol. 6(2): 31–58.
Twine, R., (2007), 'Thinking across species – a critical bioethics approach to enhancement', *Theoretical Medicine and Bioethics*, Vol. 28(6): 509–523.
Whatmore, S., (2002), *Hybrid Geographies – Natures, Cultures, Spaces*, London: Sage.
Wolfe, C., (ed.), (2003), *Zoontologies: The Question of the Animal*, London: University of Minnesota Press.
Wolfe, C., (2003), *Animal Rites – American Culture, the Discourse of Species, and Posthumanist Theory*, London: University of Chicago Press.
Womack, J.E., (2005), 'Advances in Livestock Genomics: Opening the barn door', *Genome Research*, Vol. 15: 1699–1705.
Wu, C.H., K. Nomura, T. Goldammer, T. Hadfield, B.P. Dalrymple, S. McWilliwam, J.F. Maddox, J.E. Womack and N.E. Cockett, (2008), 'A high-resolution comparative radiation hybrid map of ovine chromosomal regions that are homologous to human chromosome 6 (HSA6)', *Animal Genetics*, Vol. 39: 459–567.
Young, I.M., (1990), *Justice and the Politics of Difference*, Princeton: Princeton University Press.

Life times

Tim Newton

Introduction

This chapter will explore the possibility of differences in our perception of the 'times of nature', especially that relating to our biological and social selves. 'Nature' is of course an ambiguous term (Soper, 1995; Keller, 2006, 2008). Its invocation can as easily evoke images of nature as the 'repressed' (the 'rape' of nature, pollution, climate change, etc) or the 'repressed' (as when 'nature' is used to assert the 'natural' superiority of Caucasians or men). In addition, nature evades classification since labels such as the 'biological', 'physical' and 'social' aspects of nature refer to zones which lack clear boundaries (since they are closely intertwined).

Yet in spite of the ambiguity of 'nature', I shall argue that we still need to refer to distinctions within nature since it does not encompass a uniform terrain. Such contention questions our approach to the sociology of biology and genomics. In particular, it asks whether we can treat the social and the biological/genomic in equivalent terms. In this chapter, I will suggest that it is problematic to make this assumption because there remain certain differences in our perception of the biological, physical and social domains of nature, particularly with regard to their temporality.

In part, this argument will be pursued by examining the project of 'molecular anthropology'. Attention will be paid to the concept of the 'molecular clock' and to the popularisation of notions of 'genetic ancestry'. It will be argued that notions such as the molecular clock conflate social, biological and natural temporality. The molecular clock concept aims to devise a biological 'constant' capable of measuring human history over millennia and between species. As such, it raises the question of whether we can transpose images of temporality between the physical, biological and social sciences.

This kind of conflation is not however unique to molecular anthropology. It can also be a feature of social science discourse, especially that following from anti-dualist contention (such as Adam, 1988, 1995, 1998; Urry, 2000). In consequence, following an initial outline of molecular anthropology, I shall explore some of the limitations of anti-dualist argument within sociology and the social sciences. The discussion will then broaden to consider other aspects of biological and social temporality as well as the issue of transhumanism.

In advancing some of the arguments in this chapter, I am aware that they run counter to an almost 'metanarratival' desire for anti-dualism within some work on nature, including that relating to genomics, the body, or the natural environment (Newton, 2003a, 2003b, 2007). Yet the problem with an overweening anti-dualism is that it obscures the possibility of difference in our perception of nature.

Molecular anthropology

As Marianne Sommer notes, the term 'molecular anthropology' was coined in 1962 at a conference which aimed 'to arrive at a better integration of anthropology into the evolutionary synthesis' (2008: 478). Through a variety of developments, it came to apply aspects of genomics to concerns within biological and evolutionary anthropology. These included the comparison of 'human and nonhuman primate populations', patterns of 'genetic variation in various contemporary human populations', 'genetic variation in nonhuman primate populations', and the 'retrieval of genetic information directly from ancient specimens' (Stoneking, 1997: 87). By comparing the similarities and differences between humans and nonhuman primates, it was assumed that we could learn how both sets of primates have developed and evolved. At the same time, molecular anthropologists might discover 'the history in our genes' (Relethford, 2003: 1) and thereby move beyond the limitations of conventional social history or traditional historical artefacts. The temporal perspective could become long term, millennial in its interest. As the anthropologist, John Relethford, noted, 'this view of history deals with more than just written history or recent times. The focus here is on the evolutionary history of a population, or a group of populations, or even an entire species' (2003: 4).

In part, these developments reflected the way in which 'biological anthropologists, for the most part, became more attached to developments coming from the biomedical sciences' (Goodman and Leatherman, 1998: 15). For some, this meant working 'alongside geneticists, cell biologists, and anatomists, rather than with linguists, archaeologists, and cultural anthropologists' (Goodman and Leatherman, 1998: 15). In part, it also reflected long standing concerns within the discipline such as that with blood lines and kinship since, for many early anthropologists, 'blood . . . meant . . . genetics and biology' (Schneider, 1972: 32).

As Barnes and Dupré note, 'genomes can now be described as four-letter sequences billions of letters long, and this has made unimaginably large numbers of new relations of similarity and difference accessible to us' (2008: 98). Not surprisingly therefore, 'bioinformatic and computational techniques' are argued to be 'key to . . . studies' in molecular anthropology (Disotell and Fiore, 2008: 20). In describing their molecular anthropology 'laboratory' at New York University, Disotell and Fiore note that 'internet connectivity . . . and access to high performance computing resources are as important as laboratory bench space and instrumentation' (2008: 20). By automating the comparison of non-recombinant DNA sequences (principally in portions of mitochondrial DNA and the Y-chromosome), molecular anthropologists hope to determine the similarities and differences that have evolved over millennia within human populations, and between humans and other primates. A high level of genomic similarity is generally interpreted as indicative of close kinship or close primate relations on the basis that 'given certain assumptions, measures of genetic similarity can be taken as proportional to the length of time species [or groups] shared a common ancestor' (Relethford, 2003: 33).

The popularisation of such techniques has facilitated a range of books aimed largely at the general reader, with titles such as '*Mapping Human History: Genes, Race, and Our Common Origins*; *The Great Human Diasporas: The History of Diversity and Evolution*; *Genes, Memes and Human History: Darwinian Archaeology and Cultural Evolution*' (Bivins, 2008: 17). In addition, molecular anthropological technology promises new ways for individuals to 'find', 'discover' and 'know' themselves (Rose, 1990, 2006) by determining their 'ancestral past', and a number of genetic 'testing' companies have arisen to cater for this developing need (such as '*Genetic Genealogy*').

Yet when popularised, the dissemination of molecular anthropological argument can further misleading impressions, as in the argument that differences between humans and chimpanzees are minimal because we share '98 per cent' of our DNA. As one advocate turned critic has argued, such contention creates an exaggerated sense that humans and chimpanzees are close relatives (Marks, 2002: cf., Barnes and Dupré, 2008). This is a difficult proposition to maintain when humans and chimps do not appear to be alike. In other words, there remains a strong case for human exceptionalism, in spite of attempts within social science and popular media to anthropomorphize other primates (Meyer, 1999; Newton, 2007). Human beings remain strongly distinctive from other primates if only because we are 'bipedal and culture-reliant' (Marks, 2002: 44) and exhibit exceptional technological flair (Burkitt, 1999; Newton, 2007). Though molecular anthropologists accept that 'humans possess culture and apes do not' (Relethford, 1990: 248), they have tended to downplay the differences between humans and other primates (eg Relethford, 1990). Nevertheless asserting human difference does not mean that humans are 'superior': we can espouse human exceptionalism without advancing human supremicism (cf., Twine, this volume).

However some research suggests that there is as much as a 99.4 per cent genomic similarity in the coding DNA between humans and chimpanzees (Wildman *et al.*, 2003, cited in Barnes and Dupré, 2008: 103). If this is the case, and if the genome is to constitute a primary influence on human morphology and behaviour, the remaining 0.6 per cent needs to play a very large part in accounting for the notable differences between humans and chimps. Given that human evolution seems likely to reflect an interpolation of genomic mutation, selection, environment (eg microbial), and intergenerational cultural communication (Burkitt, 1999; Newton, 2007), a more compelling explanation is that the genome is not quite as significant to human behaviour as many geneticists, molecular biologists and molecular anthropologists still like to believe.

Molecular anthropologists may counter such critique by noting that their field is concerned with far more than the genetic similarity between humans and chimpanzees, and may, say, include the study of the genomes of ancient microbial systems, or the evolution of shared health risks in primates and humans. Nevertheless, molecular anthropology does appear open to some fairly conventional critique. For instance, critics have pointed to the cultural and linguistic assumptions that inform molecular anthropology. As Marks notes, the determination and classification of similarities within human populations, or between and humans and other primates, can only remain 'a cultural act' (2002: 48). Or as Barnes and Dupré argue, our 'ways of classifying things or situations are deeply entrenched in our linguistic practices' (2008: 96). Assertions 'that humans and chimps are very alike' reflect existing cultural and linguistic assumptions (Barnes and Dupré, 2008: 102). At the same time, the analysis of genome sequences is necessarily ambiguous and open to varied interpretation. As Barnes and Dupré further observe, 'there are innumerable methods of making such comparisons, each liable to yield a different result' (2008: 102).

In asserting that scientific analysis is culturally and linguistically relative, the project of molecular anthropology evokes debate familiar to sociologists, such as that between realism and constructionism (as well as its precursors), and their application to discussion of the interrelation between social science and biology (eg Murphy, 1994; Soper, 1995; Barnes, Bloor and Henry, 1996; Dickens, 1996; Bury, 1997; Burningham and Cooper, 1999; Macnaghten and Urry, 1998; Hacking, 1999; Irwin, 2001; Newton, 2003a, 2007). For example, in common with variants of realist argument elsewhere in the social sciences, molecular anthropologists appear likely to respond to linguistic and constructionist critique by arguing that, though analysis is necessarily circumscribed by the ambiguities of language, this does not amount to a dismissal of molecular anthropology *en tout*.

Yet at the same time, debates between realists and constructionists can be seen to pivot around their differing perceptions of temporality. For example, realists tend to argue against the historical specificity of the social arena, and exhibit a much stronger belief in the existence and long term endurance of social structures than that generally witnessed among constructionists (Abbott, 2004; Newton, 2003a, 2007). In this light, it is interesting to note that some of the key limitations of molecular anthropology are highlighted by the temporal concepts which the field employs, particularly that of the 'molecular clock'.

In what follows, I shall question the images of temporality engendered by this concept. In addition, I will examine the popularisation of molecular anthropological technique through commercial attempts to determine an individual's long-term ancestry.

Yet before further exploring these issues, it is firstly necessary to take a detour. This is because the assumption that temporality can be unproblematically treated across the social and natural domain is not unique to molecular anthropology. Instead, it can be observed within sociology and social science, partly as a result of the predominance of anti-dualist critique. In consequence, I shall firstly attend to anti-dualist contention and then explore wider aspects of biological and social temporality. Following this review, its argument will be applied back to the consideration of the concept of the molecular clock, as well as the popularisation of molecular anthropological technique that follows from attention to the 'ancestral genome'.

Beyond anti-dualism

Dualism has frequently been presented as the evil 'sin' to which any right minded social thinker should object. For example, Grosz asserts that:

> Dualism . . . is . . . at least indirectly responsible for the historical separation of the natural sciences from the social sciences and humanities, the separation of physiology from psychology, of quantitative analysis from qualitative analysis, and the privileging of mathematics and physics as ideal models of the goals and aspirations of knowledges of all types. Dualism, in short, is responsible for the modern forms of elevation of consciousness . . . above corporeality (1994: 7)

In such formulations dualism carries a heavy load, being responsible for many of the ills of knowledge construction since the Enlightenment. As John Meyer (1999) notes, critique of dualism has represented a common tactic in the social science treatment of nature where Western thought is frequently portrayed as co-extensive with the 'cleavage between humanity and nature' (Meyer, 1999: 7). Yet as Meyer (1999) further observes, it is naïve to assert this dualistic tradition if only because of the commonality of *anti-dualist* sentiment within Western thought. According to Meyer, this has included the likes of 'St Francis of Assissi, Spinoza, Emerson, Thoreau, Kropotkin and Heidegger' (Meyer, 1999: 10). Meyer asserts that such writers cannot be discounted as a 'minority tradition', and in consequence, we need 'call into question . . . the very existence of a dominant dualistic paradigm in the first place' (1999: 10). In addition, as Meyer argues:

> If Western thought were truly so dualistic, after all, then presumably most of us (those influenced by the history of Western thought) should find the view that humans are constituted by nature to be ultimately *incomprehensible*, rather than more or less incontrovertible. That this position is comprehensible and even compelling suggests that it is not as novel or as threatening to extant thought as many imagine' (1999: 11, original emphasis)

In a similar vein, Meyer suggests that critique of the social science treatment of nature has often overplayed the extent to which 'negative' conceptions have held sway. For example, writers such as Mathews (1991), Adam (1998) and Urry

(2000) have decried the social science reliance on Newtonian assumptions about nature on the grounds that they incorporate mechanistic and atomist thought. These writers all argue that social science has been dominated by an outmoded Newtonian model of temporality and natural science that needs to be replaced by post-Newtonian proposition, such as that relating to quantum mechanics and complexity theory. Adam argues that the influence of a Newtonian thought on social science has been highly detrimental because Newtonian argument 'excludes life . . . knowledge . . . any kind of human activity, emotion, interest and frailty' (1998: 40) and is 'dominated by Cartesian dualism' (1990: 152).

Yet as Meyer notes, for this kind critique to work, it is important to assume that 'only one conception of nature [is] . . . dominant in any particular era' (Meyer, 1999: 16). For example, if Newtonianism is merely 'an interpretation of some particular thinkers within the West, rather than the . . . dominant model' (Meyer, 1999: 15), it cannot be seen as the evil culprit which every right minded thinker must reject. In other words, the critique of Mathews (1991), Adam (1998) and Urry (2000) tends to assume that its evil targets, such as Newtonian or Cartesian thought, have hitherto represented relatively all encompassing and pervasive philosophies which are in need of urgent replacement by more anti-dualist and holistic accounts of nature.

This argument has a number of limitations. First, the argument that an epoch is dominated by a single, or singular, model of nature is too simplistic since any particular historical period tends to be characterised by competing philosophies. Secondly, the outright rejection of models of nature, such as those of Newtonian and Cartesian thought, is too draconian since an intellectual vandalism may result if we reject certain writers outright (for example, see Levinas's interesting use of Cartesian conjecture; Levinas, 1994). Thirdly, anti-dualist critique tends to downplay possible differences between 'human society' and the rest of nature. Yet as Meyer argues, we can accept human difference and exceptionalism without rejecting 'nature's evident role in constituting who we are' (1999: 19; Catton and Dunlap, 1980).

In sum, the desire to break down the barriers between nature and culture, mind and body, the social body and the biological body, 'natural time' and 'social time', psychology and physiology, etc, represents a laudable project. Yet it is also carries the danger of denying the possibility that there can be differences between the biological, physical and social aspects of nature. At its most fervent, it implies that having demolished the barriers, we can move beyond 'bypassing strategies' (Latour, 1999: 17) and build a common ontology and epistemology across the natural and social sciences. Yet can we defend such a single and singular perspective, a theory to encompass everything, if not 'of everything'? Is it really the case that 'the difference between "nature" and "cult ure" . . . glides into a zone of indiscernibility' (Connolly, 2002: 62)? What if some differences remain in our understanding of 'natural' and 'social' process?

In particular, can we treat 'biological', 'physical' and 'social' temporality in equivalent terms? This is an important issue because the assumption that we can collapse 'natural' and 'social' temporality can be commonly observed amongst both life and social scientists. For instance, in her introduction to a text on chronobiology, Patricia DeCoursey describes the development of time measurement techniques, from the reading of the heavens to water, mechanical and electric clocks. She then draws parallels between these '*physical*' measurements and the development of techniques to assess 'internal biological clocks' (2004: 9). Examples of these '*biological*' assessments include ancient Greek observations of the 'daily rhythms of leaves opening and closing', 18th century French studies of circadian rhythms, and the more detailed 20th century understanding of circadian biology. In describing these developments, DeCoursey draws further parallels with human '*social*' development by, for example, noting the 'importance of the mysterious forces of sun, moon, planet, and stars in the everyday and religious life of primitive peoples' (2004: 5).

Throughout DeCoursey's discussion it appears that physical, biological and social temporality are equivalent. In comparing across these provinces, it seems that we are on a similar terrain and that the temporal processes described are essentially the same. Yet the question remains as whether we can apply such parallels between the times of nature in an unproblematic manner. In the remainder of this chapter, I shall pursue this issue and then consider its implications for the construction of temporality within molecular anthropology.

Different temporalities

In *Nature and Sociology* (Newton, 2007), I suggested that although the 'natural' and the 'social' are completely interwoven, this does *not* mean that we perceive the natural and social domain in exactly equivalent terms (Elias, 1991a, 1992). In particular, I argued that there remain differences in our perception of 'natural' and 'social' temporality. On the one hand, as writers such as Adam (1995) have argued, our sense of social time is directly related to our observation of nature since we rely, for example, on the rotation of the earth to tell us the time of day. On the other hand, social temporality appears characterised by a remarkable plasticity of process that does not seem to be exactly mirrored in our perception of other times of nature.

To explore this issue further, the ensuing discussion will firstly consider images of temporality within biological process, and then contrast them with those pertaining within the social arena. In discussing biological temporality, I shall take life and natural science knowledge 'seriously'. In other words, I do *not* accept the argument that because natural science knowledge is inevitably socially mediated, we are rendered entirely incapable of adjudicating as to its likely veracity, or that we must remain 'neutral' and 'bracket out' its findings (for further discussion, see Soper, 1995; Burningham and Cooper, 1999; Hacking, 1999; Irwin, 2001; Newton, 2007).

Images of biological temporality

In *Nature and* Sociology I argued that, in temporal terms, many biological processes contain elements of what might be called a 'hyper-dynamic' *stability* of process. To take a quotidian example, human beings appear to have been characterised for millennia by features such as two eyes, ears, arms, legs etc, and this stability of outcome suggests some remarkable *stability* of process. A similar longevity in morphology is of course apparent in many other species. As Evelyn Fox Keller observes, 'in each generation the fertilized egg grows, with astonishing dependability, into an adult that is still clearly recognizable as a member of that species' (2000: 105). Although this process may be as much 'developmental' as 'genetic', there remains an extraordinary fidelity in our processes of inheritance, and fossil remains suggest that this fidelity has been maintained over several millennia.

Yet as many writers on genomics have noted, this kind of apparent stability is reliant on cellular processes which are far from stable, as is illustrated by the complex interactions which surround DNA. As Keller notes, 'DNA is not intrinsically stable: its integrity is maintained by a panoply of proteins involved in forestalling or repairing copying mistakes, spontaneous breakage, and other kinds of damage in the process of replication' (2000: 26–7). In other words, its fidelity seems to be the *product* of remarkable biological dynamism. On the one hand, 'our bodies are made of extraordinary unstable material' (Cannon, 1932: 37, cited in Birke, 1999: 96), and yet on the other, except when we experience ill-health, we *consistently* reproduce them. We do not just continually 'shed our skin' but consistently recreate all our organs and our 'blood and bones'. Biologically we appear to be characterised by images of constancy and phenomenal flux.

This hyper-dynamic stability also appears to apply to other biological processes. For example, evolutionary biologists indicate that aspects of cellular process, such as the existence of a membrane and cytosol, represent a biological ordering of considerable duration (Davey, Halliday and Hirst, 2001). On the one hand, certain common evolutionary changes can be very fast and dynamic, as with bacteriology and virology. On the other hand, some of the *processes* involved in rapid evolutionary change seem to show a notable longevity and stability. For example, in spite of their 'almost unlimited mutation mechanisms' (Dupré, 2006: 13), viruses appear to be composed of either double or single stranded DNA or RNA, elements which seem to be characterised by a remarkable duration. Similarly, this argument is not negated if one accepts 'neo-Lamarckian' arguments that evolution proceeds partly through epigenetic inheritance since the latter still appears reliant on processes that are probably of considerable duration, such as those surrounding DNA methylation or 'genomic imprinting' (Jablonka and Lamb, 1999: 96–112). Indeed, the debate about epigenetic inheritance partly involves a quest for 'mechanisms' of epigenetic inheritance, or an '*epigenetic inheritance system*' (Jablonka and Lamb, 1999: 80, original emphasis) that would explain the *long-term* duration of neo-

Lamarckian processes *across time*. It might of course be objected that the relationship between nucleotide sequences, cells and proteins has become far more confusing, ambiguous and uncertain. Yet there still seems to be agreement that there are such things are RNA 'ribbons' or strands of DNA, even though the 'gene' represents a concept that has long been 'in trouble' (Keller, 2000: 66). In other words, although there are significant doubts concerning key molecular biology concepts, and their sometimes cavalier application, it still appears that the processes surrounding DNA, RNA, mRNA, RNAi, etc, have remarkable longevity and duration within genetic and epigenetic inheritance. In addition, in spite of the instability of genomic processes, there remains remarkable fidelity in many of their 'outcomes'. Otherwise, it would not be the case that most of us have two hands and two eyes, and two eyebrows, ears, lips, feet, legs, arms, etc, and that all of these are of an identifiably human form. It is in the sense that we seem to see a hyper-dynamic stability of process.

Images of social temporality

What interests me is that this image of temporality is not so readily apparent in that which we conventionally call the 'social', where an inter-millennial stability of process or outcome appears unlikely. Even the more pressing cases for social constancy appear 'fleeting' when viewed from the millennial timescapes which we associate with some physical and biological processes. For instance, we may think of our emotions as 'natural', or human beauty as 'timeless', or capitalism as the 'end of history', even though our emotions are culturally and historically conditioned (Lutz, 1988; Elias, 1994; Newton, 1998), human beauty was once perceived as Rubenesque rather than slim, and capitalism is a very recent development. From a millennial perspective, seeming 'natural' human constants appear transient. Similarly, we like to think of other human concerns as 'timeless', such as the questions posed by religion or death. Yet we forget that in 1000 years time, the *form* of religious practice – as distinct from that of, say, carbon, water or DNA – is likely to have changed considerably, or in strongly secular societies, it may even have disappeared altogether. Similarly, the meaning of death is subject to change, as witnessed in the transition from the medieval experience of a 'public' death to its present privatisation 'behind the scenes of normal life' (Elias, 1985: 85), where it is sequestered or denied (Giddens, 1991; Bauman, 1992; Mellor and Shilling, 1993; Willmott, 2000). On the one hand, there are certain features which can appear part of the 'human condition' such as religion, social inequality, or ethnic conflict. On the other, these human 'characteristics' continually change. For example, the popularity of religions ebb and flow, the terms of ethnic conflict gradually evolve, and the definition and proportion of those in poverty does not remain stable. In general, the meaning of supposed human constants such as emotion, beauty, religion, conflict, poverty and death remain mutable and subject to social transformation. In no meaningful sense can they be said to exhibit the longevity of biological elements such as DNA and RNA, or physio-chemical compounds such as carbon, hydrogen and oxygen[1].

In sum, it remains difficult to formulate stability in human social affairs. Yet as a consequence, it is also difficult to completely collapse the times of 'nature' and 'society', or apply the same sense of temporal performance to the natural and social sciences[2]. This is because the social arena appears to be associated with far greater historical specificity than that which we see in the rest of nature. It may of course be fairly objected that this argument needs to be qualified, especially with regard to our biology. As George Gaylord Simpson (1967) argued, historical sciences such as evolutionary biology must necessarily address the uniqueness and contingency of historical events (cf., Sommer, 2008). Yet as noted above, this contingency does not mean that we lack longevity in our biological selves, whether it is the constancy of having two arms, legs, etc., or the persistence of RNA ribbons and DNA strands. In other words, although the biological domain is characterised by historical contingency, and phenomenal 'speed' and flux, it can also evince a longevity of process that is not similarly matched within the social realm. In sum, the *historical longevity* of certain biological processes and outcomes presents a marked contrast to the *historical specificity* of the social domain. Biological and social temporality is therefore not simply equivalent even though we exist within a bio-social terrain that remains nothing if not intimately intertwined.

Transhumanism

It might be objected that the above arguments are invalidated by the kind of bio-futures anticipated by variants of posthumanism (see Twine, this volume). This is particularly the case with the technological determinism exhibited by transhumanists who believe that we will soon be able to redesign the human body according to cultural and consumer desire (eg Stock, 2002). In these bio-futures, our biology becomes as changeable as our social existence, and this lack of biological fixity means that 'biological time' may collapse into 'social time'. In other words, both 'body and society' will become 're-programmable', plastic and changeable, and therefore highly historically specific.

Yet the problem with this argument is that it overlooks the significance of natural constancies for human biosocial functioning. To put this another way, it remains unlikely that nature and culture will be 'equally subjected to re-imagination' (van Dijck, 1998: 177) because people will probably want to retain a fair amount of constancy in human biology. For example, would we ever want 'unreliable' bodies? Even if were to create a completely new 'transhumanist' body, would we not still desire constancy? When our bodies lack constancy of process, the results for the individual can be serious ill health, and much medicine is concerned with trying to 'fix' such bodily inconstancy. Medical technique desires constancy by 'stabilising' complaints such as coronary arrhythmia or a hyperactive thyroid.

It is of course possible to envisage alternate sci-fi worlds where our bodies might change according to the task at hand (van Dijck, 1998). For instance, perhaps our eyesight would be able to 'switch levels' and suddenly see things at a microscopic level, or our fingers might become smaller and longer so as to deal with some intricate feat of dexterity. This would represent bodily plasticity nearly on a par with that of social language. Yet this does not mean that biological constancy and replicability would go out of (cultural) fashion. We would *not* want to suddenly go blind when we 'switched' to 'microscopic perception mode' or find that our fingers had turned into 'elongated pincers that had become incapable of grasping anything' (Deleuze and Guattari, 1988: 72). In other words, constancy is likely to remain a *socially desirable* feature of our biology even if it could be subject to redesign, or endlessly adaptable to the task in hand. The same feature is not a *necessary* characteristic of our social lives. Instead, the open-ended flexibility of the human social arena means that social stability and constancy is something that must be continually performed. Most significantly, plasticity in the social arena does not equate with 'dysfunctionality'. In contrast, an unfettered biological plasticity implies impairment since our bodies would cease to maintain their 'health'.

In sum, even if we ignore the reductionism of transhumanist bio-futures, and their assumption that the 're-engineering' of biological life can be reduced to genetic manipulation, it still remains highly doubtful that we would ever want to collapse biological time into social time in any kind of posthuman universe, at least in the sense that biology might one day become as changeable as our social universe.

In what follows, I will extend the foregoing argument by further consideration of what we mean by human history, based on an initial comparison between the tropes of social history and the attention to 'molecular anthropology' provided by bio-anthropologists and evolutionary biologists (see above). In so doing, I wish to argue that the *longevity* of some natural processes does *not* necessarily equate to an ability to *specify constancy* across time. In other words, just because we see some conditional regularities in some physical and biological process, it does not follow that we can use natural processes to 'measure' 'evolutionary' time.

Tales of human history

To the extent that historians are reliant on human symbolisation (Elias, 1991b), they must confront the ambiguities, slippages, dissemblings, contradictions, etc that form an everyday part of spoken, written and mathematical communication. In consequence, it is conventional for historians to try to compare between different sources pertaining to any historical period and thereby discern the variation and veracity of different historical accounts. To this extent, social history is a necessarily '*messy*' affair where it is difficult to construct a foundational account of changing human habits, patterns and sensibilities. In other words, following conventional hermeneutics, the indeterminacy and ambiguity of human language and symbolisation means that it is difficult to posit a single 'true', 'objective' or 'scientific' account of human history.

In this context, it is interesting to return to molecular anthropology and its attempt to cut across the 'messiness' of human history. By drawing on work on protein and DNA sequencing, molecular anthropology promises a more objective account of human history based on the belief that 'the closer to the gene the more reliable was the analysis' (Sommer, 2008: 478). 'Genetic' analyses could thereby overcome the indeterminacy of human social affairs and the limitations of conventional social history.

Of particular interest is the deployment of the concept of the 'molecular clock' by biological and molecular anthropologists (Dietrich, 1998; Sommer, 2008). This concept drew on the biochemical expertise of researchers such as Emile Zuckerkandl, Linus Pauling, Allan Wilson and Vincent Sarich. Conceptually, the notion of the molecular clock assumed that 'the rate of mutation was roughly constant' across species (Dietrich, 1998: 92), or more specifically that 'the observed changes in the amino acid sequences of proteins from different species should be "approximately proportional in number to evolutionary time" (Zuckerkandl and Pauling, 1965: 148)' (Dietrich, 1998: 99). Even though the rate of evolution at the level of the organism was assumed to be 'highly variable', it was argued that at 'the molecular level [it] was constant' (Dietrich, 1998: 86). In consequence, one could devise a '*molecular time scale for human evolution*' (Wilson and Sarich, 1969: 1088, added emphasis). The molecular clock would provide a biological constant capable of assessing human history across evolutionary time. Human history might no longer represent a messy and indeterminate terrain.

Yet there are several problems with the molecular clock concept (Dietrich, 1998; Sommer, 2008). For instance, there remains the 'paradox' of its assumption that 'molecular evolution proceeds in a rather regular fashion with respect to time' whereas 'organismal evolution is classically considered to be an irregular process' (Wallace, Maxson and Wilson, 1971: 3129, cited in Dietrich, 1998: 108). As Barnes and Dupré note, 'in order to measure evolutionary time, molecular clocks need to tick with random mutations, but the *clocks are actually part of what is evolving* and evolution is reckoned to involve the non-random process of natural selection' (2008: 114, added emphasis). In addition, it has been argued that linear and constant 'clock-like' change is unlikely to occur due to 'erratic overdispersion', or the process by which 'proteins evolve erratically through time and across lineages' (Rodriguez-Trelles, Tarrio and Ayala, 2001: 11405). Rodriguez-Trelles et al argue that evidence of erratic overdispersion amounts 'to a denial of there being a molecular clock' (2001: 11410).

Such difficulties in maintaining the concept of the molecular clock are not all together surprising if one considers the extraordinary ambition which it represents. The attraction of the molecular clock to molecular anthropologists derives from its supposed regularity, and its consequent ability to act as a 'benchmark' by which to derive constancy (in spite of the plasticity and uncertainty of human biosocial development). Indeed, the idea of the molecular clock constitutes a desire for the kind of constancy associated with physical processes (Elias, 1992). Yet the question arises as to why notions of constancy should be transplanted to the biological arena in this fashion. In effect, the molecular clock hypothesis aims to devise a biological constant capable of measuring history over millennia and between species. This represents an astonishing aspiration based on the translation into the biological sphere of the kind of constancy we associate with physical constants such as gravity, Planck's constant and the speed of light.

The example of the molecular clock reinforces the question of whether we can transpose images of temporality from the physical sciences to the biological sciences, or elsewhere within nature[3]. It also confronts the relationship between longevity and constancy of *measurement*. In this chapter and elsewhere (Newton, 2007), I have argued that some natural processes can be characterised by greater longevity than the historical specificity associated with the social arena, as is exampled in the life sciences by the longevity of DNA, RNA, etc. However this does not mean that the *longevity of natural process 'A' can necessarily provide a constant measure for evolutionary characteristic 'B'*, as is desired by the concept of the molecular clock. Yet this appears to be the effective desire of molecular anthropology in it's ambition to arrive at a consistent measure of the messiness of human history based on an appeal to a supposedly constant and regular process.

This project is likely to fail because, (1) the molecular clock constant does not appear consonant with that associated with aspects of physics, and (2) its intermillenial ambition is based on an unwarranted assumption that biological time, as reflected in the 'molecular clock', can be made equivalent to social time and the plasticity and indeterminacy of human history. It seems to assume that time itself is a kind of constant measure that can applied equally, and unproblematically, across the physical, biological and social domains. Yet as has been argued above, there remain differences in our perception of the times of 'nature' and 'society' and, in spite of anti-dualist desires, we cannot just collapse biological and social time as though they are the same 'unity' – any more than we can collapse our physical and biological senses of time.

Popular molecularization

In spite of the limitations of the 'molecularisation' of human history, it has received increasing attention in popular texts and media. For example, Roberta Bivins cites a headline from *The Guardian* in 2005, which confidently asserts that 'all of human history can be written with four letters' (2008: 15). Such media reports further the image of 'molecular history' as a 'new' field of study that can provide a seemingly '"scientific", objective, and especially, a *complete* history' (Bivins, 2008: 16, original emphasis). As illustration, Bivins cites the publicity material for a recent book (Cavalli-Sforza's *Genes, Peoples and Languages*):

> Historians relying on written records can tell us nothing about the 99.9 per cent of human evolution which preceded the invention of writing. It is the study of genetic variation, backed up by language and archaeology, which provides *concrete evidence about the spread of cultural innovation, the movements of peoples . . . the precise links between races*. (Bivins, 2008: 16, added emphasis).

As Bivins further observes, 'we see the gene, or clusters of genes acting as a new kind of historical evidence, for a new kind of history, a "history" with periodization often closer to that of geology than of social or political history' (2008: 17).

Yet the problem with this popular molecularisation of human history is that it assumes that our 'ancient' biological selves can somehow be equated with our social and cultural history. In other words, it once again presupposes that we can collapse biological and social time and fuse them into an 'artificially linear history' (Bivins, 2008: 19). Though these assumptions remain problematic, there does nevertheless appear to be a popular desire to 'read' our 'ancestral selves' through our genome. This is reflected in the services of a growing range of genetic 'testing' companies such as *Oxford Ancestors*, *DNA Tribes, Genetic Genealogy, African Ancestry and AfricanDNA* (Bivins, 2008). For example, *Oxford Ancestors* promises the ability to determine 'the personal link between you and your ancestral clan mother', and whether you are descended from 'the High Kings of Ireland' or perhaps from 'Somerled, the 12th century Celtic hero who drove the Norse from the Isles' (Oxford Ancestors, 2009). Furthermore if you purchase a book by the founder of *Oxford Ancestors*, Bryan Sykes (Professor of Human Genetics at the University of Oxford), you can learn how 'almost all Europeans can trace their ancestry back to one of seven women', whom Sykes named '"The Seven Daughters of Eve"' (Oxford Ancestors, 2009; see Sommer 2008). *Genetic Genealogy* (2009) goes further and promises their clients the ability to determine whether their ancestry includes historical figures as varied as Thomas Jefferson, Genghis Khan, Niall of the Nine Hostages (another 'great Irish King'), The Romanovs, Jesse James, and Marie Antoinette.

Companies such as *Oxford Ancestors* do acknowledge that 'there is no genetic basis for ethnicity or race' (Oxford Ancestors, 2009), whether 'celtic' (the 'High Kings of Ireland', etc) or 'biblical' ('daughters of Eve', etc). Yet other companies such as *Genetic Genealogy* and *African Ancestry* are not so cautious. For instance, *African Ancestry* promises to be able to tell African Americans the 'specific African ethnic group' from which they are descended (African Ancestry, 2009). As commerce, companies such as *African Ancestry* and Henry Louis Gates' *AfricanDNA* trade on an understandable desire of some African Americans to discover a history beyond slavery. However these desires remain problematic to the extent that 'ancestral' genetic testing companies, whether *African Ancestry, Genetic Genealogy* or *Oxford Ancestors*, rely on an admixture of romanticism and reductionism. Firstly, they rely on a romanticisation of history, especially where their clients hope for a 'blood line' that links them to some mythical or distinguished ancestry, such as 'Somerled, the 12th century Celtic hero' (Oxford Ancestors, 2009), 'Marie Antoinette' (Genetic Genealogy, 2009) or the 'Yoruba of Nigeria' ('testimonials', African Ancestry, 2009). Secondly, they tend toward a 'Genes R Us' reductionism based on 'a worldview in which human diversity is increasingly ascribed to genetic causality' (Taussig, Rapp and Heath, 2003: 61). Such assumptions remain problematic if only because it is difficult to attribute causal responsibility exclusively to 'genes'. As Jonathan Kaplan comments, 'while a gene may be associated with a particular variation on a trait in one environment, it may not be associated with it in another' (2000: 47–48). As a consequence, it is impossible to know quite what the inheritance of some 'ancestral DNA' can tell us about either ourselves, our ancestors, or our supposed present day 'relatives' living in some far off land.

In effect, companies such as *Oxford Ancestors, Genetic Genealogy, African Ancestors* or *DNA Tribes* may tell us more about the contemporary cultural desire to 'psychologise' ourselves than they do about the socio-cultural concomitants of our individual ancestry. Instead of using contemporary psychological technique to 'understand who we are' (Rose, 1990), 'ancestral' genetic testing trades on a growing desire to 'discover' ourselves by analysis of our ancestral past. For example, a UK television programme entitled *Who do You think You Are* (BBC, 2009) encourages celebrities to 'emotionally relate' to the trials, tribulations and achievements of their ancestors. Through this process, the celebs come to (often tearfully) know themselves, not only through the conventional psychotherapeutic interest in the immediate family, but also by coming to understand the lives of several generations of their forebears. Genetic testing companies *go further* and promise 'genetic ancestral knowledge' across centuries or even millennia. As such, they proffer a new kind of 'genealogical psychology' (Foucault, 1979) that promises an ability to know oneself through genetic determination of one's ancient past. For instance, rather than taking a psychological test (of, say, 'personality'), individuals can now take a 'DNA test' and thereby find out 'who they *really* are'. Yet such 'genetic' additions to 'psy culture' (Rose, 1990) may engender an even more dubious reading of ourselves than that conventionally associated with psychological testing (Hollway. 1989; Newton, 1994). By proffering a supposedly scientific reading, they may further spurious accounts that falsely assume that culture and history can be reduced to the 'truth' of the genome.

In sum, the question remains as to whether we can cut across the messy social complexity of human history, whether through ancestral genetic analysis or through appeals to the kind of biological constants implied by the 'molecular clock' or the 'anthropological gene'. These projects appear deficient to the extent that they try to collapse biological and social time, and transpose images of linearity and constancy onto a human social history characterised by plasticity and uncertainty. Analysis of the genome does not imply greater reliability or validity, whether we wish to understand the social context of our ancestors or devise biological 'constants' capable of measuring biosocial time across several millennia.

Conclusion

The desire to relate natural and social temporality can appear as a laudable project. Most commonly, it reflects an anti-dualist concern to move beyond the Cartesian 'divide' and accept that 'nature' and 'society' are fundamentally 'one'. And to a very large extent, there is nothing wrong with this supposition. The problem arrives when one assumes that natural process is therefore equivalent to social process, or that 'biological', 'physical' and 'social' temporality can be collapsed on the grounds that they are 'one and the same'. Though this assumption is highly questionable, it can be observed in a diverse range of conjecture, from the transhumanist futurism of *Humanity+*, to the anti-dualism of writers such as Mathews (1991), Adam (1995) and Urry (2000), to the endeavours of 'molecular' anthropologists to 'measure' evolutionary history through a supposed biological clock, or the commerce of companies that promise to reveal our ancestral socio-cultural history. Collectively, all these projects assume that 'social' time can be either collapsed or easily correlated to 'biological' time, either by making our bodies as plastic as our societies, or by using supposed molecular biological constants to 'cut across' the 'messiness' of human history, or by drawing metaphorical parallels between 'natural' and 'social' time (eg Urry, 2000), or by invoking romanticised accounts of our ancestral socio-cultural past through an illusory appeal to the 'certainty' of the genome.

The question which applies to all these diverse endeavours is that of whether one can interweave images of 'biological' and 'social' temporality, or use the former to calibrate the latter. Their collective problem is that they tend to forget the possibility of difference in our perception of temporality across physical, biological and social spheres. In this sense, a problem with an overweening anti-dualism is its tendency to obscure the possibility of difference. Even though human beings constitute a social and biochemical 'unity of substance' (Spinoza, 1996), this does not mean that we perceive temporality across the domains of nature in exactly the same way.

Does this mean that there are clear differences in kind (Hacking, 1999) and a semantic dualism between discourses of the 'natural' and the 'social' (Changeux and Ricoeur, 2000)? Possibly, but we should still fight shy of resurrecting dualisms between 'nature' and 'society', or 'brain' and 'mind' (see Newton, 2007). What may be more appropriate to say is that perceived differences in natural and social temporality make it difficult to apply uniform modes of inquiry across these domains, whether through futurist 'transhumanist' images of bio-plasticity, the collapsing of natural and social time, or the desire of a molecular anthropology and ancestral genomic analysis to cut across the messiness of human affairs.

Acknowledgments

Thanks to Sarah Parry, Richard Twine and John Dupré for their helpful comments on earlier drafts of this chapter.

Notes

1. Social instability is as much a characteristic of 'process' as 'outcome'. For example, capitalism is sometimes cited as a longue durée social process (Braudel, 1985). As a consequence, Andrew Collier suggests that we can ascribe conditional regularity to capitalist process over lengthy periods of time:

   > [social] laws can be formulated in terms which are universal, by virtue of being *conditional*: 'if the ownership of productive wealth is separated from the direct producers and divided between competing sellers of the products, then tendencies x, y, z will operate' (Collier, 1994: 244, original emphasis).

   The problem however with such proposition is that it is difficult to make a convincing argument that capitalism constitutes a longue durée process. In particular, a strong argument can be made that capitalism has varied considerably over the relatively recent past. As I argued elsewhere (Newton, 2003c), the early systems of financial credit employed in the English sixteenth century are fundamentally different from those found in present day cash economies. In addition, social relationships within capitalism can quickly change even in the short term. For instance, the interdependency between employer and employee, or between landlord and tenant, can change rapidly through industrial relations or housing legislation, or through variation in the extent to which such legislation is locally enacted. To take a British example, employer-employee relations changed rapidly during the UK Thatcher administration of the 1980s. In consequence, it is difficult to speak of the kind of conditional social regularities that realists such as Collier desire (see above), whether over the short or long term. To put this another way, there remains a good deal of validity to Collingwood's assertion that:

   > Specific types of human organization, the city-state, the feudal system, representative government, capitalist industry, *are characteristic of certain historical ages* (Collingwood, 1946: 211)

2. In part, one can of course argue that these temporal images of the social and the biological are just a question of scale. For instance, Ted Benton acknowledged that 'long term historical prediction' is difficult in the social sciences because of the temporal variance of its subject matter (1981: 18). Benton then went on to argue that substantially the same situation applies in the natural sciences: 'epistemologically speaking, the situation is quite comparable with the natural sciences. On the very much greater time-scale of biological, geological and cosmological change the comparable long-term historical prediction is equally suspect' (1981: 18). Yet this seems a strange comparison. To say that physical and biological science breaks down and is 'suspect' over millions of years can hardly be used as a justification for the failure of the social sciences to predict well even in the extremely short term. Many physical and biological processes appear to us to exhibit durability over the very long term, a situation which does not seem to apply within the social arena.

3. As Williard Koukhari and Robert Sothern note, the drawing of 'analogies between biological clocks and mechanistic models [i.e. physical temporality] is a temptation far too common' (2006: 89), and one might argue, frequently inappropriate.

References

Abbott, A., (2004), *Methods of Discovery: Heuristics for the Social Sciences*. New York: W.W. Norton.

Adam, B., (1988), Social versus natural time: A traditional distinction re-examined. In Michael Young and Tom Schuller, (eds), *The Rhythms of Society*. London: Routledge: 198–226.

Adam, B., (1990), *Time & Social Theory*, Cambridge: Polity.

Adam, B., (1995), *Timewatch: The Social Analysis of Time*, Cambridge: Polity.

Adam, B., (1998), *Timescapes of Modernity: The Environment and Invisible Hazards*, London: Routledge.

African Ancestry, (2009), Available at: http://www.africanancestry.com/benefits.html. (last accessed 13 Jan 2009).

BBC, (2009), *Who Do You Think You Are?* Available at: http://www.bbc.co.uk/programmes/b007t575 (last accessed 14th October 2009),

Barnes, B., D. Bloor and J. Henry, (1996), *Scientific Knowledge: A Sociological Analysis*, London: Athlone.

Barnes, B. and J. Dupré, (2008), *Genomes and What to Make of Them*, Chicago: University of Chicago Press.

Bauman, Z., (1992), *Mortality, Immortality and Other Life Strategies*, Cambridge: Polity.

Benton, T., (1981), 'Realism and social science: Some comments on Roy Bhaskar's "The Possibility of Naturalism"', *Radical Philosophy*, No. 27: 13–21.

Birke, L., (1999), *Feminism and the Biological Body*, Edinburgh: Edinburgh University Press.

Bivins, R., (2008), 'Hybrid vigour? Genes, genomics, and history', *Genomics, Society and Policy*, 4 (1): 12–22.

Braudel, F., (1985), *Civilisation and Capitalism, 15th to 18th Century, Vol. 3: The Perspective of the World*, London: Fontana.

Burkitt, I., (1999) *Bodies of Thought; Embodiment, Identity and Modernity*. London: Sage.

Burningham, K. and G. Cooper, (1999), Being constructive: Social constructionism and the environment, *Sociology*, 33 (2): 297–316.

Bury, M. R., (1997), *Health and Illness in a Changing Society*, London: Routledge.

Cannon, W. B., (1932), *The Wisdom of the Body*, London: Kegan Paul, Trench and Trübner.

Catton, W. R. and R. E. Dunlap, (1980), 'A new ecological paradigm for post-exuberant sociology', *American Behavioral Scientist*, 24 (1): 15–47.

Cavalli-Sforza, L. L., (2001), *Genes, Peoples and Languages*, Berkeley: University of California Press.

Changeux, J-P. and P. Ricoeur, (2000), *What Makes Us Think? A Neuroscientist and a Philosopher Argue About Ethics, Human Nature, and the Brain*, Princeton: Princeton University Press.

Collier, A., (1994), *Critical Realism: An Introduction to Roy Bhaskar's Philosophy*. London: Verso.

Collingwood, R. G., (1946), *The Idea of History*. Oxford: Clarendon Press.

Connolly, W. E., (2002), *Neuropolitics: Thinking, Culture, Speed*. Minneapolis: University of Minnesota Press.

Davey, B., T. Halliday and M. Hirst, (2001), *Human Biology and Health: An Evolutionary Approach*, 3rd edn., Buckingham: Open University Press.

DeCoursey, P. J., (2004), 'Overview of biological timing from unicells to humans', in Dunlap, J.C., J. J. Loros and P. J. DeCoursey (eds), *Chronobiology: Biological Timekeeping*, Sunderland, Mass.: Sinauer Associates: 3–24.

Deleuze G. and F. Guattari, (1988), *A Thousand Plateaus: Capitalism and Schizophrenia*, London: Athlone Press.

Dickens, P., (1996), *Reconstructing Nature: Alienation, Emancipation and the Division of Labour*, London: Routledge.

Dietrich, M. R., (1998), 'Paradox and persuasion: Negotiating the place of molecular evolution within evolutionary biology', *Journal of the History of Biology*, 31: 85–111.

Dijck, J. van, (1998), *Imagenation: Popular images of genetics*, New York: New York University Press.

Disotell, T. and A. D. Fiore, (2008), 'Bioinformatics in molecular anthropology', *Connect: Information Technology at NYU*, Spring/Summer 2008: 20–23.

Dupré, J., (2006), 'The Constituents of Life: Spinoza Lectures 1 and 2, University of Amsterdam', paper delivered at an *Egenis* seminar (ESRC Centre for Genomics in Society), University of Exeter, 28th November, 2006.

Elias, N., (1985), *The Loneliness of Dying*, Oxford: Blackwell.

Elias, N., (1991a), *The Society of Individuals*, Oxford: Blackwell.

Elias, N., (1991b), *The Symbol Theory*, London: Sage.

Elias, N., (1992), *Time: An Essay*, Oxford: Blackwell.

Elias, N., (1994), *The Civilizing Process*. Oxford: Blackwell.

Foucault, M., (1979), *The History of Sexuality, Vol. 1*. London: Allen Lane.

Genetic Genealogy, (2009), Available at: http://www.dnaancestryproject.com/ydna_intro_famous.php?id=marieantoinette&typ=m (last accessed 18 June 2009).

Giddens, A., (1991), *Modernity and Self-Identity: Self and Society in the Late Modern Age*, Cambridge: Polity.

Goodman, A. H. and T. L. Leatherman, (1998), 'Traversing the chasm between biology and culture: An introduction', in Goodman, A. H. and T. L. Leatherman (eds), *Building a New Biocultural Synthesis*, Ann Arbor: University of Michigan Press: 3–41.

Grosz, E., (1994), *Volatile Bodies: Toward a Corporeal Feminism*, Indianapolis: Indiana University Press.

Hacking, I., (1999), *The Social Construction of What?* London: Harvard University Press.

Hollway, W., (1989), *Work Psychology and Organisational Behaviour*. London: Sage.

Humanity+, (2010), 'Transhumanist Declaration' available at: http://humanityplus.org/learn/philosophy/transhumanist-declaration/ last accessed 6 May 2010.

Irwin, A., (2001), *Sociology and the Environment: A Critical Introduction to Society, Nature and Knowledge*, Cambridge: Polity.

Jablonka, E. and M. J. Lamb, (1999), *Epigenetic Inheritance and Evolution: The Lamarckian Dimension*. Oxford: Oxford University Press.

Kaplan, J. M., (2000), *The limits and lies of human genetic research: Dangers for social policy*, New York: Routledge.

Keller, E. F., (2000), *The century of the gene*, Cambridge, Mass.: Harvard University Press.

Keller, E. F., (2006), 'What's in a word? Genes, heritable traits and heritability', paper delivered at the University of Exeter, 13th December 2006.

Keller, E. F., (2008), 'Lecture: Nature and the natural', *BioSocieties*, 3: 117–124.

Koukhari, W. L. and R. B. Sothern, (2006), *Introducing Biological Rhythms*, New York: Springer.

Latour, B., (1999), 'On recalling ANT; in John Law and John Hassard', *Actor Network Theory and After*, Oxford: Blackwell/The Sociological Review: 15–25.

Levinas, E., (1994), *Totality and Infinity: An Essay on Exteriority*, translated by A. Lingis. Pittsburgh: Duquesne University Press.

Lutz C. A., (1988), *Unnatural emotions: everyday sentiments on a Micronesian atoll and their challenge to western theory*, Chicago: University of Chicago Press.

Macnaghten, P. and J. Urry, (1998), *Contesting Natures*, London: Sage.

Mathews, F., (1991), *The Ecological Self*, London: Routledge.

Marks, J., (2002), *What it Means to be 98% Chimpanzee*, Berkley, LA: University of California Press.

Mayr, E., (1961), 'Cause and effect in biology', *Science*, 134: 1501–1506.
Mellor, P. A. and C. Shilling, (1993), 'Modernity, self-identity and the sequestration of death', *Sociology*, 27 (3): 411–431.
Meyer, J. M., (1999), 'Interpreting nature and politics in the history of western thought: The environmentalist challenge', *Environmental Politics*, 8 (2): 1–23.
Murphy, R., (1994), *Rationality and Nature: A Sociological Inquiry into a Changing Relationship*, Boulder, Colorado: Westview Press.
Newton, T., (1994), 'Discourse and agency: The example of personnel psychology and 'Assessment Centres'', *Organization Studies*, 15 (6): 879–902.
Newton, T., (1998), 'An historical sociology of emotion?' in Bendelow G. and S.Williams (eds), *Emotions in Social Life: Social Theories and Contemporary Issues*. London: Sage: 60–80.
Newton, T., (2003a), 'Crossing the great divide: Time, nature and the social', *Sociology*, 37 (3): 433–457.
Newton, T., (2003b), 'Truly embodied sociology: Marrying the social and the biological?' *The Sociological Review*, 51 (1): 20–42.
Newton, T., (2003c), 'Credit and civilization', *British Journal of Sociology*, 54 (3): 347–371.
Newton, T., (2007), *Nature and Sociology*, London: Routledge.
Oxford Ancestors, (2009), Available at: http://www.oxfordancestors.com/component/option,com_frontpage/Itemid,1/(last accessed 13 January 2009).
Relethford, J. H., (1990), *The Human Species: An Introduction to Biological Anthropology*, Mountain View, Calif.: Mayfield.
Relethford, J. H., (2003), *Reflections of Our Past. How Human History is Revealed in Our Genes*, Boulder: Westview.
Rodriguez-Trelles F., R. Tarrio and F. J. Ayala, (2001), 'Erratic overdispersion of three molecular clocks: GPDH, SOD, and XDH', *Proceedings of the National Academy of Sciences*, 98 (20): 11405–10.
Rose, N., (1990), *Governing the Soul*, London: Routledge.
Rose, N., (2006), *The Politics of Life Itself: Biomedicine, Power, and Subjectivity in the Twenty-First Century*, Princeton, NJ: Princeton University Press.
Schneider, D. M., (1972), 'What is kinship all about?' in Reining, P. (ed.), *Kinship Studies in the Morgan Centennial Year*, Washington, D.C.: The Anthropological Society of Washington: 32–63.
Simpson, G. G., (1967), *The Meaning of Evolution*, revised edition, New Haven: Yale University Press.
Sommer, M., (2008), 'History in the gene: Negotiations between molecular and organismal anthropology', *Journal of the History of Biology*, 41: 473–528.
Soper, K., (1995), *What is Nature? Culture, Politics and the Non-Human*. Oxford: Blackwell.
Spinoza, Baruch/Benedict de, (1996), *Ethics*, by Curley, E. (ed. and trans.) and Hampshire, S. (Introduction). Harmondsworth: Penguin.
Stock, G., (2002), *Re-designing Humans: Our Inevitable Genetic Future*, Boston: Houghton Mifflin.
Stoneking, M., (1997), 'The human genome project and molecular anthropology', *Genome Research*, 7: 87–91.
Taussig, K-S., R. Rapp and D. Heath, (2003), 'Flexible eugenics: Technologies of the self in the age of genetics', in Goodman, A., D. Heath and S. Lindee (eds), *Genetic Nature/Culture*, Berkeley: University of California Press: 58–76.
Urry, J., (2000), *Sociology Beyond Societies: Mobilities for the Twenty-First Century*, London: Routledge.
Wallace, D. G., L. R. Maxson and A. Wilson, (1971), 'Albumin evolution in frogs: A test of the evolutionary clock hypothesis', *Proceedings of the National Academy of Sciences*, 68: 3127–3129.
Wildman, D. E., M. Uddin, G. Liu, L. I. Grossman and M. Goodman, (2003), 'Implications of natural selection in shaping 99.4% nonsynonymous DNA identity between humans and chimpanzees: Enlarging genus *Homo*', *Proceedings of the National Academy of Sciences*, 100: 7181–7188.

Willmott, H., (2000), 'Death. So what? Sociology, sequestration and emancipation', *The Sociological Review*, 48 (4): 649–665.

Wilson, A. and V. Sarich, (1969), 'A molecular time scale for human evolution', *Proceedings of the National Academy of Sciences*, 63: 1088–1093.

Zuckerkandl, E. and L. Pauling, (1965), 'Evolutionary divergence and convergence in proteins', in Bryson, V. and H. Vogel (eds), *Evolving Genes and Proteins*, New York: Academic Press: 97–166.

Afterword

Barry Barnes

A love of order and ordering is ubiquitous among humans and in modern, differentiated societies there are occupational niches that cater for all, however strong or weak that love may be. The museum curator places pieces of polished rock in sequence on plinths, each with a description on an accompanying card, and thereby one of many orderings, perhaps the temporal order in which geological strata were laid down, is displayed. No matter how well defined the bounds of these strata are and how consistent their layering from one place to another; the plinths and cards will express a natural order. Indeed it is hard to imagine anything more satisfying than minerals as objects for ordering and sorting. The ordering of the stars may be intrinsically satisfying, but our hands cannot grasp them and arrange them and cast away anomalies. And whilst human artefacts are also good to order, they can pose awkward challenges and require recognition of untimely truths. As to living things, they offer endless opportunities, to the laboratory scientist as well as the curator and the taxidermist. But we are perhaps too interested in them, so that their classification attracts more than its share of controversy, and the very latest technoscientific innovations, like those associated with the rise of genomics, may be utilized so rapidly and frenetically to reinforce or renew their ordering that they can appear to be having the opposite effect.

This book owes much to those who sort and order and to those who create the means of doing so. If one would celebrate disorder, then order is vital. To set things out of place, they must have a place. And the book is indeed a valorization of disorder and disordering, naturalized here and there as belief in a messy and unruly nature. Perhaps the desire to create disorder is another human universal: perhaps the pleasure of the toddler, removing the bottom piece from the swaying tower of dominoes she has trained a parent to build and rebuild, is also one that all humans share. But beneath the tensions that this should lead us to expect with ordering and sorting of any kind, other more particular and pressing conflicts and concerns are almost invariably to be found.

All the contributions to the book problematize natural order in some way or other, whether through the dissolution of its presumed objects, or their hybridization and admixture, or the erosion of their boundaries and divisions, or the removal of what protects them from threatening externalities, whether unnatural, or non-natural, or super-natural. And many subsidiary themes in the book invite understanding as but special cases of this one dominating theme. When things are ordered or classified, multitudes of objects are placed into just a few groups or kinds and a simplification is achieved often reckoned helpful to understanding and to action. Accordingly, the collection celebrates complexity: the experience presented to our understanding is multifaceted and our understanding of it should be nuanced – a term of which the authors are especially fond. Similarly, order facilitates successful prediction, and here the spotlight turns from predictions to promises and to the frequent failures to fulfil them we have now learned to expect. And whilst natural order is widely reckoned to be incarnate in the world itself, open to inspection, here the emphasis is on that order as a human creation, the problematic outcome of a process of naturalization, often cobbled together in processes of co-production wherein boundaries that are at once social and epistemic are comprehensively demolished.

How are we to understand this sustained critical reflection on the creation of natural order? Surely not as an attempt to enlighten the orderers and sorters themselves. They will have encountered most of the problematic objects and anomalies cited as challenges already overcome, by inventing new categories in an existing scheme of things perhaps, or moving the objects from one place to another therein, or reinterpreting rules of classification, or rejecting the objects as pathological. In other words, alleged sources of disorder and taxonomic disruption are likely to be known on the other side as familiar kinds of nuisance, or even as the very things that make their work a source of genuine satisfaction and not just dreary routine. We might even wonder whether some creators of natural order do not have their own specific methods of classifying those who would throw things back into chaos once more and bring to nought the efforts of those who over many generations have laboured at the construction of natural order, or 'nature' as it is now sometimes called.

Fortunately, it is the work of a moment to confirm that the authors of this collection are not solely concerned with the creation of chaos and the celebration of disorder. Their critical reflections are for the most part directed to more positive ends. Nor do they confine themselves, in the manner of radical social constructivists, to nature in the sense of the taxonomies and systems of classification created by humans. Nature, the editors tell us, is in one crucial sense 'the very stuff that makes up our physical world; it is a material presence'. As such it is a 'locus of classification systems for ordering the world', systems which allow knowledge of nature to be formulated verbally and representations of nature to be constructed in linguistic form, creating the natural order as a cultural product that is nature in a second crucial sense. This allows the question to be raised of whether some of the cultural products that constitute nature in this second sense may not be better or worse than others when considered as versions of nature in the first sense. And indeed a variety of answers to this question are to be found here, including repeated claims that nuanced versions of nature, that stress its complexity and unruliness, are to be preferred to those that impose simplicity, and the more specific claim that there are superior alternatives to 'genocentrism' as the basis for a classification of living things.

If the book has a weakness it is not that it holds back from constructive positive assertions but that it offers less than it might by way of justification for them. And here I have sympathy with the authors; for I both share some of their preferences and recognize a lack of fully convincing arguments on their behalf. I am at one with them, for example, in taking a contestable materialist view of the natural world. It has the pragmatic advantage of sustaining the old convention whereby our words were assumed to refer to matter independent of us, located in the external world – whether a bit of it, like the moon, or the lot, nature. Where this default was overridden and speech acts were used to refer to themselves or elements of themselves the change was signalled with inverted commas, as 'moon' for the notion, or 'nature' for the presumed order of nature entire and the whole array of notions constituting that. A contrast of this sort serves to sensitize us to two different ways in which natural objects may be said to be reconstituted or transformed. One way involves a transformation in 'nature', as when a human embryo becomes a pre-embryo to be succeeded by an embryo, or billions of dandelions become four, or a sex change becomes a gender change, or Oedipus is recognized as his mother's husband and his daughter's brother. The other, generally more interesting to biologists, involves a transformation in nature, as when a bomb explodes or a bird being experimented upon is given bird flu or a promised Mycoplasma Laboratorium is eventually fully synthesized. In the first kind of transformation nature just lies there and gets spoken of, as it were, whilst in the second nature acts or is acted upon.

Of course, both kinds of transformation can occur at once, and are liable to be found entangled together as part of the larger 'entanglement' that the editors discuss. The bombs that were dropped on Iraq destroyed museum collections and books and people and thereby changed both nature and 'nature' together. But it is helpful nonetheless to be able to speak of 'both'. Here is one reason why a chapter on synthetic biology has an important role in this collection. Synthetic biologists aspire to change nature, not 'nature', although the latter would surely follow were they to claim success in any of their major aims. Certainly, we might expect accounts of 'nature' to complete and conflict for a time after the synthesis of a new life form, as the limitless resources of our discourse were set to the task of deciding whether it had really happened at all, how much it really mattered, whether it did or did not falsify the theory of evolution by natural selection, and so on and so forth. Whether synthetic biologists themselves would wish to make major contributions to debates of this sort, or whether they are too involved in trying to cure malaria and produce bio fuels, I do not know. But if they did, it would surely be on the basis of their assumption, documented in Jane Calvert's chapter, that nature is simple in itself and amenable to reductionist explanatory schemes, an assumption that stands in striking contrast to that of the book itself, with its affection for nuance and complexity and anathematization of reductionism. Here is the crosscutting theme of the book that most clearly confronts us with the question of how, from the outside, we are to decide which of different sets of ontological and epistemic preferences is the better.

Elsewhere in the collection the focus is a little more on 'nature' than on nature, that is, on classifications and narratives and theoretical interpretations. I'll discuss the authors' own discussion of these by posing two questions: Are some methods of producing these things superior to others? Are some versions of them superior to other versions? These questions are both important, but it is the first that is much the more difficult. In particular, it is hard to identify some apparent weakness in method or procedure that does not on further examination turn out to be employed very widely and difficult to manage without.

Take for example, the claim that 'genocentric' and other biologically based methods of classification focus highly selectively on just some aspects of living things and are capable of supporting only incomplete explanations of their functioning. The trouble with this as a general criticism is that no mode of classification escapes it. It is a standard point about 'material' that it is indefinitely rich in information and that 'complete' information is never available, if indeed the notion itself is meaningful. And it is worth adding that fatigue is sure to triumph before every possible way is explored of ordering and sorting the 'incomplete' information that is gathered and building explanatory accounts upon it. Sociological studies of laboratory research can frequently be read as accounts of how much of the limited amount of information that is gathered is nonetheless ignored, discounted, left for another day, denied validity or reliability, so that attention is able to focus on a minute proportion of 'relevant' findings and science is made possible. But studies of this sort need not be read as implying criticism. Those contributions to the present volume that include interview materials vividly convey the potential endlessness of the tasks of information gathering, sorting, interpretation, inference drawing and so forth, and the utter irrelevance of the notion of 'complete' description to the evaluation of any actual account. Scientists, we might say, merely make a virtue of necessity when they are selective.

Of course one should not make too much of a virtue of necessity, as can occur when the tiny amount of knowledge so far produced is proclaimed to be all the knowledge one needs – 'all that really matters' or 'everything essential' for an understanding of something, despite the lack of knowledge of the knowledge that is lacking, as it were. There is indeed a genuine critical opportunity here, but it needs to be handled with care. In her fascinating book on *Satisfying Scepticism* (2001), Ellen Spolsky comments on depictions of Doubting Thomas in early modern European art. One altarpiece has Thomas accompanied by two worthy contemporaries of the painter, one with a pair of the latest fashion in spectacles perched on the end of his nose, peering intently at the body of Jesus, evidently seeking hard for something and not finding it. Jesus stands there patiently, as if waiting for them to realize their mistake. And if it is knowledge of what is real or what is essential that they are seeking, then there surely is something mistaken about seeking it through obsession with the empirical, even though Thomas' mistake here has been repeated over and over throughout history.

Things get designated as real or essential precisely to distinguish them from what is merely empirical or empirically indicated; and throwing empirical observations at such designations risks overlooking why they are made and the work they actually do in our discourse. Minimally, essentialism is a discursive strategy used to identify some part of the cognitive and epistemic order as settled and stable, not open for negotiation at the moment as mere contingent observations or generalizations may be, not subject to adjustment in the context of science as part of the continual juggling of the components of knowledge systems and their referents that characterize the growth of a field. Similarly, a minimal realism is frequently encountered as a discursive strategy that facilitates criticism of specific empirical claims, and who would deny that criticism of that sort is needed, both in the context of everyday life and the context of science?

Minimally, in science, essentialism is a shared strategy to facilitate selectivity of a given sort, to encourage forgetfulness, and thereby to facilitate coordination among a community of researchers – a strategy implemented through assertions about the contents of the world. And in popularizations of science additional layers of essentialism (and realism) are almost invariably found, facilitating the further simplification necessary if a general reader is to find these texts intelligible. But whilst we might sometimes want to criticize scientists for excesses of forgetfulness and popularizations for employing essentialist strategies that encourage, sometimes deliberately, systematic misunderstandings of the knowledge supposedly being popularised, we need to recognize as well that nobody manages without essentialist and reifying forms of discourse and take care not to criticize these strategies *per se* – unless, that is, we wish also to be critical of ourselves and our own critical discourse. Certainly, a path to better classification will not be created by taking a machete to all the reifications and essences that stand in the way.

Moving to the discussion of particular classifications, and why some might be better or worse than others, it will be clear by now that only pragmatically based evaluations are to be looked for, not a general account of what 'being better' involves, still less an account of what 'the best' classifications of living things will consist of. Pragmatic evaluations, however, have much to be said for them and many of those to be found in this book are of great interest. Several contributors discuss classifications and narratives good for policies of various sorts, and by implication bad from the perspectives of those opposed to these policies. There are papers that could easily have borne the titles: 'The Politics of Stem Cell Classification', 'The Politics of Food Classification', and 'The Politics of Species'. In addition to these studies, wherein much of the relevant politics is immanent in the situation and the goings on therein, Sarah Parry provides a discussion of 'The Politics of Nature', showing how the resources of political philosophy may be drawn upon to provide the basis for an external evaluation of such activities. Elsewhere again, contributors confidently proceed on the basis of their own political intuitions; important among which, and rightly so, is that any classification close to being 'hegemonic' cannot be quite so good as it is cracked up to be and deserves to have some criticism thrown at it.

Yet another way of evaluating new versions of natural order is to use the standards that their creators themselves appeal to when seeking to justify them. This is an easy option and it is interesting with what restraint it is deployed here. Promissory rhetoric, for example, has long been referred to as 'hype', and the all too familiar narrative of scientific experts hyping up the methods of their fields for what look like expedient reasons and then as often as not failing to deliver can be applied to genomics as easily as any. If there is an implicit judgement here that this is no longer news, and that a more nuanced evaluation of how far fields meet the standards embodied in their own rhetoric is now appropriate, then I can only agree.

How now to end the end? Certainly not with a quick glance at nature to check out the merits of this book, although merits it clearly has. Crucially, it lives up to its own promissory rhetoric and effectively furthers the stated aims of its editors, both through argument and analysis and through the fascinating interview material that it brings together. More generally, the division of scientific labour is now so extensive and the utilitarian drive behind it is now so strong, that broad critical reflection cannot be fully developed as part and parcel of the business of research itself, and a variety of additional modes of understanding such are included here need to be deployed if scientists as well as those in the societies that host them are to grasp its full import. 'Nature' is something that scientists as creators and re-creators of order are expected to transform, but not so radically and chaotically that the collective representations that constitute it disintegrate. Even in differentiated societies like ours there is a point at which epistemic disorder becomes intolerable. Without a 'nature' on which there is at least some level of agreement, our minds would decouple from each other and lose coordination, and in so doing would cease to be minds at all – whatever minds are. This is the level at which our appreciation of what the sciences currently do on our behalf needs to be articulated. Their deep complicity in the constitution of our minds, and thus in our culture and practice entire, demands acknowledgement of their profound importance, just as it demands the most far-reading critical analysis of what they are about that current scholarship is capable of sustaining.

Reference

Spolsky, E., (2001), *Satisfying Scepticism,* Ashgate: Aldershot.

Notes on contributors

Barry Barnes is a Codirector of Egenis (ESRC Centre for Genomics in Society), University of Exeter, at which he was formerly professor of sociology. Barry is known for his pioneering work on the sociological study of knowledge generation and evaluation in science, and on the basis of the credibility of scientific expertise. He also has a longstanding interest in the fundamental problems of the social sciences, particularly in collective action problems, in status groups as generators of [exclusionary] collective action, and in self-referring knowledge as constitutive of social order and systems of power. Since the mid-1990s he has increasingly focused his work upon the new human biotechnologies and their social and cultural significance. He is the author of several books on the sociology of the sciences and was awarded the J. D. Bernal Prize for his career contribution to the field.

Jane Calvert is an RCUK Academic Fellow in the ESRC Innogen Centre, University of Edinburgh. After a BSc in Human Sciences (Sussex) and an MSc in History and Philosophy of Science (LSE), she did her doctoral work at SPRU (Sussex). She then worked at the ESRC Centre for Genomics in Society at Exeter University, where she became interested in the insights that the philosophy of biology can bring to the social study of genomics. At Edinburgh, Jane's research is in the broad area of the sociology of the life sciences, with a particular interest in both systems and synthetic biology.

Gail Davies is Senior Lecturer in Geography at University College London, UK. She is interested in the contemporary intersection of the social, the spatial, the biological and the ethical, particularly in the constitution of biological relatedness and difference, and in the relations between different ways of knowing biology. She has published in journals such as *Economy and Society*, *Environment and Planning*, *Geoforum*, *Health and Place*, *Progress in Human Geography*, *Public Understanding of Science*, and *Transactions of the Institute of British Geographers*. She is currently tracing the unfolding spaces of postgenomics and the shifting location of global science through the changing use of mice as model organisms.

John Dupré is the Director of Egenis (ESRC Centre for Genomics in Society) and Professor of Philosophy of Science, University of Exeter. His research has specialized in the philosophy of biology. His current work focuses on philosophical issues concerning the interpretation and implications of genetics and genomics. His publications include: *The Disorder of Things: Metaphysical Foundations of the Disunity of Science* (Harvard, 1993); *Human Nature and the Limits of Science* (Oxford, 2001): and most recently, with the distinguished sociologist Barry Barnes, *Genomes and What to Make of Them* (Chicago, 2008).

Nicola J. Marks is a lecturer in Science and Technology Studies at the University of Wollongong in Australia. She moved there after completing her PhD and one year of post-doctoral research at the University of Edinburgh. Her research looks into social aspects of the biosciences, particularly public engagement in stem cell research.

Richard Milne is a Research Associate in the Department of Geography at the University of Sheffield, working on consumer understandings of risk, anxiety and trust in the context of UK food scares. His PhD research, conducted at University College London, focused on the geographies and futures of science, bringing together work in the sociology of expectations with the geographies of food and technoscience. It examined how publics and experts place biopharming, the production of pharmaceuticals using genetically modified crops, in relation to agricultural and pharmaceutical biotechnologies.

Tim Newton is Professor of Organization and Society at the University of Exeter. His current research interests include sociology and 'nature' (the body, biology, genomics, the natural environment), interdisciplinarity, the historical sociology of credit and commercialization, economic sociology and the 'new' sociology of finance. His work has been published widely in journals of sociology, psychology and organisation studies. His most recent text is *Nature and Sociology* (London: Routledge, 2007).

Sarah Parry is a lecturer in Sociology within the Science, Technology and Innovation Studies subject group and Associate at the ESRC Innogen Centre, University of Edinburgh. There are two interrelated threads to her research interests. The first focuses on the social and cultural aspects of knowledge and meanings of biological material. She is currently writing a book on the relational meanings of different matter and the social and experiential contexts in which these are constructed. The second focuses on the scope for public participation in decision-making. She has developed an active interest in the role of the sociologist in society and doing 'interactive social science' with regards to public and policy framings of sustainable consumption.

Celia Roberts is a Senior Lecturer in the Department of Sociology at Lancaster University, with strong links to the Centre for Gender and Women's studies and the Centre for Science Studies. She works on issues relating to biology, reproduction and biomedicine, embodiment, feminism and innovative health technologies. She is currently involved in two European projects: one on older people and new technologies of care, and one on patient organizations. She is the author of *Messengers of Sex: Hormones, Biomedicine and Feminism* (Cambridge, 2007) and is currently writing a new book on puberty.

Karen Throsby is an associate professor in the Department of Sociology at the University of Warwick. Her research explores the intersections of gender, technology and the body across a variety of sites, including reproductive technology, obesity surgery and extreme endurance sport. She is the author of *When IVF Fails: Feminist, Infertility and the Negotiation of Normality* (Palgrave, 2004) and co-editor (with Flora Alexander) of *Gender and Interpersonal Violence: Language, Action and Representation* (Palgrave, 2008).

Richard Twine is a sociologist at Lancaster University, UK. His interests are at the nexus of animal studies, gender studies and biotechnological change. He is particularly interested in posthumanism, its salience to our understanding of intersectionality as well as the importance of human/animal relations to climate change discourse. In 2010 his first book, Animals as Biotechnology – Ethics, Sustainability and Critical Animal Studies, will be published in the Earthscan Science in Society series.

Claire Waterton is Director of the Centre for the Study of Environmental Change (CSEC) and Senior Lecturer in Environment and Society within the Department of Sociology, Lancaster University. She has been a researcher in the sociology of scientific knowledge at Lancaster since 1992, focusing on the making of environmental knowledge, and the relationship of scientific and other forms of environmental knowledge to policy making at UK, EU and global scales. She has led several major interdisciplinary research projects funded by policy and academic research bodies. Her research within the ESRC's Science and Society Programme was recognized by a scholarship from the USA's SSRC and the UK's ESRC in 2005. She is co-editor, with Bronislaw Szerszynski and Wallace Heim, of *Nature Performed: Environment, Culture and Performance* (2003, Oxford: Blackwell) and is the author of many articles in the SSK and wider environmental policy literatures.

Index